AF556814

LEWES STRING THEORY WORKSHOP

LEWES STRING THEORY WORKSHOP

Lewes, Delaware
6–27 July 1985

Editors

L CLAVELLI
A HALPRIN

Published by

World Scientific Publishing Co. Pte. Ltd.
P. O. Box 128, Farrer Road, Singapore 9128

Library of Congress Cataloging-in-Publication data is available.

LEWES STRING THEORY WORKSHOP

ISBN 9971-50-033-7

Printed in Singapore by General Printing & Publishing Services Pte Ltd.

PREFACE

Interest in the theory of superstrings as a fundamental theory of matter reached near hysterical proportions in 1985 as a consequence of the significant work of Green and Schwarz showing that such theories are anomaly free and probably finite [M. Green and J. Schwarz, *Phys. Lett.* **149B**, 117 (1984); *ibid.* **151B**, 21 (1985)]. This volume is a record of that part of the euphoric period that took place at the Lewes Center for Physics in Lewes, Delaware, July 6-27, 1985.

The workshop opened with a brief historical review of the metamorphosis of this subject from a partially successful description of hadrons-only to a viable theory of everything. It continued with a series of pedagogical talks by Paul Frampton, Michio Kaku and Louise Dolan.

More specialized topics were treated in the second and third weeks concluding on a phenomenological note with a discussion of a model for hadronic strings in nuclei.

We thank all those who attended the workshop and remained attentive amidst all of Beach Plum Island's wildlife distractions. We are especially grateful to all contributors who supplied manuscripts in a timely fashion. Finally, we thank the Center's support staff, especially Kay Hackett and Erin O'Connell for attending to infinite amounts of photocopying and other worldly needs of all participants.

Lou Clavelli
Arthur Halprin

CONTENTS

Historical Overview of Superstring Theory

Louis Clavelli

Dept. of Physics and Astronomy
University of Alabama
University, AL 35486*

and Dept. of Physics, Indiana Univ.
Bloomington, In. 47405

ABSTRACT

Superstring theory is a new approach to particle physics that shows promise of revolutionizing our view of matter. Einstein's gravity and Yang Mills gauge interactions emerge as approximations to a more fundamental picture that may avoid much of the inconsistency and arbitrariness of these earlier theories.

Historical reviews in physics are notorious for two serious faults. The first is that they tend to judge developments from a one sided, personal point of view. The second is that the author may unconsciously tend to exaggerate the importance of his or her own work. I can

---
* present address

easily avoid the latter fault by omitting any direct reference to my work. The former fault is, I believe, more intrinsic and can only be mitigated by averaging over many such personal accounts.

The present explosion of interest in superstring theories is almost unique in the history of such phenomena in that, for perhaps the first time, an older age group of physicists has a marked advantage over the new generation compensating for the latter's superior enthusiasm. This advantage is based on the fact that there was a classical era of string theory during the years 1968 to 1974 during which about 600 papers were written on the subject, laying an extensive and strong foundation for the present developments. Most of the string theorists of that heady era, though now in their forties, were then in the prime of their youth. Sociologically, it was the period of the Vietnam war. String theory in those days was an engrossing intellectual enterprise involving mathematical techniques that were unfamiliar to most of us, presaging the current confluence of modern mathematics and theoretical physics. For many it was a way of taking our minds off the reports of war. I will outline some of the major ideas of this classical period below. For the present it is sufficient to note that this era of string theory came to a relatively abrupt end in about 1974. The decline was a more or less direct result of the meteoric rise of gauge theories following the discovery of neutral currents and the subsequent successes of the Glashow-Weinberg-Salam model. By the end of 1974 there was a firmly established dogma in theoretical physics: 'All physical theories are gauge theories of point particles at the most fundamental level.'

The present workshop is part of the evidence that physics has now overthrown and gone beyond this dogma, but the years from 1974 to 1984 were the dark ages of string theory. There was in fact a continuing low level of activity throughout these years but, for the most part, it had become reckless to work in the field without a very secure position or a very strong patron. As a consequence very few graduate students of the past 10 to 12 years were acquainted with basic string techniques. Instead they became well versed in point particle field theory which in itself is a worthwhile subject of study. However the warts of point particle theories including gauge theories had been with us so long that physicists had become inured to their ugliness. The worst of these blemishes, which may now be cured by superstring theory, were

1) Ultraviolet infinities. All field theories of point particles that are even remotely realistic have ultraviolet infinities in them. Even when these infinities are "renormalizable", their mathematical consistency is unclear. Infinities at asymptotic momentum

transfer, even if renormalizable, imply that physical processes at extremely short distance scales are outside the scope of these theories.

Quantum electrodynamics and Electroweak theory were so successful at available energy scales that the old canard about renormalization being a process of "sweeping infinities under the rug" was taken as a joke instead of a serious complaint. It was always clear of course that, if elementary particles were extended objects, these infinities might disappear but, before superstring theory, no totally consistent picture of elementary particles as extended objects existed.

2) Non renormalizable divergences. With the inclusion of quantum gravity interacting with matter, the divergences referred to above become non renormalizable. That is, in order to have finite results, the elementary Lagrangian must have an infinite number of free parameters thus losing all predictive power. By going to a supergravity theory, these problems are postponed to higher order in the perturbation expansion but no reason exists to expect them to be totally avoided. Superstring theory is the only current prospect for a finite theory including gravity.

3) Free dimensionless parameters. The standard gauge theory of elementary interactions has almost two dozen free dimensionless parameters. A free dimensional

parameter is not undesirable since it just represents a definition of units. If a pure number appears in physics, however, its value is not just a choice of units but something that should be predicted by the theory. Physicists have agonized over some of these apparently arbitrary dimensionless quantities for years. Examples are the value of the fine structure constant and the number of charged leptons. At least one, the number of space time dimensions, has seemed such a basic datum of nature that few have searched for a deeper understanding.

In superstring theory there is apparently only one free parameter and it has dimensions of length. The dimension of space time is a critical integer away from which the theory loses all consistency. The absense of infinities requires a near unique internal symmetry group. Even the basic coupling constant appears to be predictable in the theory. Much work, of course, remains to be done to extract quantitative predictions but the theories appear to determine all dimensionless quantities.

In addition to their superiority over point particle theories in these three important aspects, Superstring theory automatically contains in itself as a low energy approximation Einsteinian gravity and Yang Mills gauge interactions thus indicating that we have penetrated to a more fundamental starting point.

It is tragic in a sense that P.A.M. Dirac, who spent

so many decades reminding us that theories with infinities are fundamentally unacceptable and that the value of the fine structure constant required an explanation, should have died within months of the resurgence of string theory.

The roots of string theory date back to the introduction by G. Veneziano[1] in 1968 of a dual, Regge behaved, crossing symmetric model for a four particle strong interaction scattering amplitude. The model underwent rapid development and generalization but the origin of string theory as opposed to that of dual resonance models can be traced to the 1970 observation of Yoichiro Nambu[2] that the generalized Veneziano model could be interpreted physically as the interaction of particles that were not pointlike but had an extension in one dimension. That is, the states of the Veneziano model were relativistic strings. In analogy with the action of a relativistic point particle being its reparameterization invariant path length, Nambu proposed a relativistic string action given by the reparameterization invariant area swept out by the string.

$$S = \frac{1}{2\pi\alpha'} \int d\sigma d\tau \, [(x' \cdot \dot{x})^2 - \dot{x}^2 x'^2]^{1/2}$$

$$= \int \frac{d\sigma d\tau [-\det\gamma]^{1/2}}{2\pi\alpha'} \qquad (1)$$

The action is described by an integral over the proper time τ of the string and a spacelike variable, σ, parameterizing lengths along the string. $x_\mu(\sigma,\tau)$ is the coordinate vector pointing from the origin to a particular point (σ,τ) on the world sheet of the string. $\dot{x}_\mu$ represents the derivative of x_μ with respect to τ and x_μ' the derivative with respect to σ. γ is the two dimensional metric matrix on the sheet. As can be seen from eq. 1 it has diagonal entries $\dot{x}^2$ and x'^2 and identical off diagonal entries $\dot{x}\cdot x'$. The reparameterization invariance consists of the independence of the action under the transformations

$$\tau \to \tilde{\tau} = f_\tau(\sigma, \tau) \tag{2}$$

$$\sigma \to \tilde{\sigma} = f_\sigma(\sigma, \tau) \tag{3}$$

This invariance through its various subgroups leads to the duality property of string amplitudes, the elimination of negative norm states, and, ultimately, to the appearance of Yang Mills and Pomeron-type interactions. The action of equation 1 was proposed as a theory of strong interactions. Although the spectrum of states was never satisfactory for hadrons, the string models of strong interactions were extraordinarily successful in reflecting the dominant features of hadron scattering such as (to mention only a few) duality, Regge behavior, transverse momentum cut off, and Pomeron properties. For his proposal of the string action, following his 1965 proposal of the SU(3) gauge theory of strong interactions (QCD), Nambu deserves credit for two of the three major ideas of strong interaction theory.

In fact the Copenhagen lectures in which Nambu's work appears were never orally presented since, on the first leg of his trip from California to Copenhagen, his car engine melted in the desert southwest and the rest of the trip was consequently cancelled.

Minimizing the action of eq. 1 leads to the equation of motion

$$\left(\frac{\partial}{\partial\tau}\frac{\partial}{\partial \dot{x}_\mu} + \frac{\partial}{\partial\sigma}\frac{\partial}{\partial x_\mu{}'}\right)[\det\gamma] = 0 \tag{4}$$

and to an edge condition at the string endpoints $\sigma = 0, \pi$.

$$\frac{\partial}{\partial x_\mu{}'}[\det\gamma]\Big|_{\sigma = 0,\pi} = 0 \tag{5}$$

If one chooses an orthogonal coordinate system, ie. one satisfying

$$\dot{x} \cdot x' = 0 \tag{6}$$

and

$$\dot{x}^2 + x'^2 = 0 \tag{7}$$

the Euler Lagrange equations 4 and 5 to reduce to the wave equation

$$\ddot{x}_\mu - x_\mu'' = 0 \qquad (8)$$

and the constraint that the σ derivative of x_μ vanishes at the ends of the string.

$$x_\mu'(\sigma=0,\ \tau) = x_\mu'(\sigma=\pi,\ \tau) = 0. \qquad (9)$$

The periodic solution to eqs. 8 and 9 is

$$x_\mu(\sigma,\ \tau) = q_{0_\mu} + p_{0_\mu}\tau$$

$$+\sum_{n=1}^{\infty} \frac{\cos n\sigma}{\sqrt{n}}\left(a_\mu(n)e^{-in\tau} + a_\mu^*(n)e^{in\tau}\right) \qquad (10)$$

In eq 10 a factor of $\sqrt{n}$ is factored out of the Fourier coefficients a_μ, a_μ^* for later convenience. Eqs 6 and 7 become sum rules for the coefficients:

$$\frac{1}{2}\sum_{n=-\infty}^{\infty} \alpha_\mu(n)\,\alpha_\mu(-N-n) = 0 \qquad -\infty < N < \infty \qquad (11)$$

where one puts

$$\alpha_\mu^0 = p_\mu \quad \text{and} \quad \alpha_\mu(-n) = -i\sqrt{n}\,a_\mu(n) \qquad n>0 \qquad (12)$$

The string is quantized by letting the Fourier

coefficients become creation and annihilation operators

$$a_\mu(n) \rightarrow a_\mu^n \tag{13}$$

$$a_\mu^*(n) \rightarrow a_\mu^{n\dagger} \tag{14}$$

satisfying canonical commutation relations

$$[a_\mu^n, a_\nu^{m\dagger}] = g_{\mu\nu}\delta_{mn} \tag{15}$$

$$[q_{0_\mu}, p_{0_\nu}] = i\, g_{\mu\nu}. \tag{16}$$

The sum rules (11) can then no longer be identically satisfied but must be enforced as constraints on the physical states of the string

$$L_N = \frac{1}{2} \sum_{n=-\infty}^{\infty} :\alpha_\mu^n \; \alpha_\mu^{-N-n}: \tag{17}$$

$$\langle\psi_{phys}| L_N - \alpha_0 \delta_{N,0} |\Psi_{phys}\rangle = 0. \tag{18}$$

The L_N constraints comprise the Virasoro algebra[3]

$$[L_N, L_M] = (M-N)\, L_{M+N} - c(N^3 - N)\delta_{N+M,0}. \tag{19}$$

Commuting with the string coordinate, the L_N satisfy

$$[L_N,\ X_\mu(\sigma,\tau)]$$

$$= e^{-iN\tau}\ [\cos N\sigma\ \frac{d}{d\tau} - i\ \sin N\sigma\ \frac{d}{d\sigma}]\ \ X_\mu(\sigma\ \ ,\tau). \qquad (20)$$

The L_N's therefore preserve the edge condition and at the string endpoints generate reparameterizations of τ alone.

Particle emission in string theories is defined in terms of a vertex operator which, in the case of the basic (Veneziano model) string ground state is

$$V_0(k) = \int d\tau :e^{i\sqrt{2\alpha'}k\cdot x(0,\tau)}: . \qquad (21)$$

If a string state satisfying the gauge constraints (18) emits a ground state string via the vertex of eq 21 the resulting string state will also satisfy the constraint 18 provided

$$[L_N, V_0(k)] = 0. \qquad (22)$$

By direct commutation one finds

$$[L_N, V_0(k)] = N(1-\alpha'k^2) \int d\tau e^{-iN\tau} :e^{i\sqrt{2\alpha'}k\cdot X(0,)}:. \qquad (23)$$

Therefore consistency is obtained only if

$$1 - \alpha'k^2 = 0. \qquad (24)$$

Since the spin zero state in a Regge spectrum satisfies

$$\alpha_0 - \alpha' k^2 = 0, \tag{25}$$

The Virasoro condition for elimination of unphysical states in the interacting theory is

$$\alpha_0 = 1. \tag{26}$$

The leading Regge trajectory has unit intercept. Because of this, the spin one state in the spectrum is massless

$$\alpha_0 - \alpha' k^2 = 1 \Rightarrow k^2 = 0 \tag{27}$$

The Virasoro algebra has an SU(1,1) or SL(2,R) subalgebra generated by the L_0, L_{+1}, L_{-1}. The duality property of the Veneziano model was seen to be related to the covariance of the particle emission vertices under this algebra which was intensively studied at MIT, at Fermilab, and at Berkeley among, perhaps, other places in 1970 and 1971. The result was a group theoretical prescription for the construction of new dual models based on the representation theory of SU(1,1). One such model, the Neveu-Schwarz model(4), was a direct ancestor of the current superstring theories. It contained half integrally moded, anti commuting fields

$$H_\mu(z) \equiv \sum_{n=0}^{\infty} (b_\mu^n z^{-n-1/2} + b_\mu^{n\dagger} z^{n+1/2}). \qquad (28)$$

The ground state emission vertex in this model was

$$V_0(k) = g \int d\tau \, :e^{ik\cdot X(0,\tau)} k\cdot H(e^{i\tau}):. \qquad (29)$$

Like the Veneziano model, and in fact like all of the new dual models that were based on the SU(1,1) scheme, the Neveu Schwarz spectrum contained a massless vector meson.

The concept of supersymmetry in particle theory was also an outgrowth of the group theoretical study of string theories. The first super algebra to appear in physics was written, before the Neveu Schwarz work, by Pierre Ramond[5] late in 1970 in the process of constructing a dual theory for free fermions. Ramond proposed a generalization of the Dirac γ matrices for strings

$$\Gamma_\mu(z) = \gamma_\mu + i\sqrt{2}\,\gamma_5 \sum_{n=1}^{\infty} (d_\mu^n z^{-n} + d_\mu^{n\dagger} z^n). \qquad (30)$$

The infinite dimensional Virasoro algebra became a super algebra with the addition of an infinite set of anti commuting "generators"

$$F_N = \int \frac{d\tau\, e^{-iN\tau}}{2\pi i\sqrt{2}} \Gamma_\mu(e^{i\tau}) \dot{X}_\mu(0,\tau) \qquad (31)$$

The central element F_0 was the square root of L_0 just as the Dirac operator was the square root of the Klein Gordan operator.

Unlike the Neveu Schwarz model, which followed inexorably from the group theoretical principles, the Ramond model for fermions involved a significant intuitive leap. The SU(1,1) generators in the Ramond sector did not annihilate the vacuum requiring therefore substantial effort before the fermion scattering amplitudes were constructed and their consistency was proven, although the amplitudes with a single fermion line(6) were written down as early as 1971.

Ramond's paper inspired a large industry in supersymmetric point particle field theories which thrived throughout the dark age of string theory. Physicists looked to supersymmetry for a solution to the hierarchy problem of spontaneously broken gauge theories and for a consistent approach to quantum gravity. However, it now seems likely that no problems in physics theory are in fact solved by supersymmetry separated from string theory.

The Ramond Fermions and the Neveu Schwarz model were soon recognized to be intimately connected since the annihilation of fermion anti-fermion was discovered to produce Neveu Schwarz bosons. In the scattering of N particles, the N τ integrations arising from the vertices are reduced to N-3 propagator integrals due to the 3 dimensional SU(1,1) invariance group. This factorization property into vertices and propagators allows unitarity to be perturbatively implemented in string theories. It was found that, in each of the ghost free models, unitarity was respected only in a critical space time dimension (eg 26 for the Veneziano model and 10 for the Ramond Neveu Schwarz (RNS) model.) In this critical dimension a new spectrum of factorizable states emerged with many of the properties seen experimentally in diffractive scattering of hadrons.

Unfortunately (for hadronic theory) the spin two state in this "Pomeron" spectrum, like the vector state in the Reggeon spectrum referred to earlier, appeared with mass zero. An enormous amount of futile effort was expended to find consistent models without these massless states. The failure of this effort was the main reason why string theories for hadrons have not progressed although there is sufficient experimental evidence for string behavior in hadronic scattering to believe that a solution to the problem certainly exists.

In 1974 Joel Scherk and John Schwarz[7] turned this impasse into an advantage by suggesting that strings be considered as theories for elementary particles instead of for hadrons. The Ramond fermions became quarks and leptons. The massless ρ mesons became the gauge bosons of

Glashow, Weinberg, Salam, and Nambu (before spontaneous symmetry breaking) and the massless pomeron became the graviton. Even from today's perspective this was a bold idea but in 1974, at the peak of enthusiasm for point like gauge theories of elementary particles, it appeared as a rather far fetched grasping at straws to save the string theory.

What their proposal did have to offer was a scheme in which massless Yang Mills and graviton particles appeared automatically, required by the self consistency of the string picture. It was found in fact that, not only did the gauge bosons and gravitons appear with the correct masses, but their couplings also, in the zero slope limit of the string theory, were exactly those expected in Yang Mills theory and Einstein gravity. That is, the trilinear and quartic gauge couplings and the universal coupling of the graviton to the energy momentum tensor emerged automatically. In this sense, if their proposal turns out to be indeed correct, it provides a deeper understanding of the emergence of gauge and gravitational forces. It is interesting to note that in this string picture, Yang Mills gauge interactions and Einsteinian gravity are only approximations (albeit exceedingly good ones at low energy) to the full string theory with non zero slope.

The Scherk Schwarz paper was not well received and in fact coincided with the end of the classical era of string theory. Many of us, under the pressures of the real world, turned our attention to gauge theory or lattice calculations. Advances in string theory came at a much slower rate after 1974 but progress did not stop altogether. It was realized, for instance, that if strings were to be taken as a theory of elementary processes, there was no need to keep the Neveu Schwarz tachyon pion state. With massless gauge bosons as external states, no tachyons would appear as intermediate states in tree graphs due to G parity conservation. They could then also be projected out of loop graphs in a way consistent with unitarity. The unwanted pion state would, in general, still be produced in fermion anti-fermion scattering. It was found however that by restricting the fermions to be of given chirality (e.g. left handed) the tachyon decoupled from these processes also. Furthermore, by taking the fermion fields to be real as well (Majorana fermions) the spectrum of the theory became fully supersymmetric.[8] The unrestricted model had a supersymmetry only on the two dimensional sub manifold. It is only possible to have chiral Majorana fermions in certain discrete space time dimensions among which is the critical dimension of the RNS model.

The work of ref 8 can be regarded as the birth of the first (type I) super string theory. It was ghost free and tachyon free. It contained a supersymmetric Yang Mills and fermion sector composed of open strings and a supergravity sector composed of closed strings. Although the generation of closed strings could not be avoided in a theory of open strings, if one chose the external states to be closed strings, open strings could be consistently eliminated. This led immediately to two more fully self consistent (type II) super string theories.

This work did not attract wide attention however until the present renaissance of string theory burst upon us in the late summer of 1984 with the announcement by Mike Green and John Schwarz that for a particular choice of the internal symmetry group, SO(32), the type I string theory was free of gauge anomalies[(9)] and was probably a totally finite theory.[(10)] Strictly speaking, their finiteness result was proven only for the one loop four point function. The extension of the theorem to higher N point functions at the one loop level will be presented in two contributions to this workshop.

The current explosion of interest in string theories has within the last nine months produced two more superstring string theories by an ingeneous hybridization of the 26 dimensional Veneziano model with the 10 dimensional restricted RNS model.[(11)] This later approach seems to offer more promise for a phenomenologically realistic theory of quarks, gauge bosons, and gravitons. Present activity in string theory would be more difficult to review in one hour than the past 15 years of string history. Progress is being made on many fronts including higher order calculations, string field theory, and string phenomenology.

I apologize to the many string theorists whose work I have not explicitly mentioned. Even if I have correctly identified the key ideas, there are certainly dozens of others who have contributed substantially to the present happy state of string theory.

This review was prepared while the author was a visitor at Indiana University supported in part by the DOE.

References

1. G. Veneziano Nuovo Cim 57A 190 (1968)
2. Y. Nambu, Copenhagen Summer Institute Lectures (1970) (unpublished)
3. M. A. Virasoro Phys. Rev. D1 2933 (1970)
4. A. Neveu and J. H. Schwarz Nucl. Phys. B31 86 (1971)
5. P. Ramond, Phys. Rev. D3 2415 (1971)
6. A. Neveu and J. H. Schwarz, Phys. Rev. D4 1109 (1971)

   C. B. Thorn Phys. Rev. D4 1112 (1971)
7. J. Scherk and J. H. Schwarz, Nucl. Phys. B81 118 (1974)
8. F. Gliozzi, J. Scherk, and D. Olive, Phys. Lett. 65B 282 (1976)

   F. Gliozzi, J. Scherk, and D. Olive, Nucl. Phys. B122 253 (1977)
9. M. B. Green and J. H. Schwarz, Phys. Lett. 149B 1175 (1984)
10. M. B. Green and J. H. Schwarz, Phys. Lett 151B 21 (1985)
11. D. J. Gross, J. A. Harvey, E. Martinic, and R. Rohm, Phys. Rev. Lett. 54 502 (1985)

INTRODUCTION TO SUPERSTRINGS

P. H. Frampton
Institute of Field Physics, Department of Physics and Astronomy
University of North Carolina, Chapel Hill, NC 27514, USA

The organizers Lou Clavelli and Arthur Halprin kindly invited me to give introductory superstring lectures at the Lewes Center for Physics, repeating those I gave recently in Trieste.

These lectures are in three sections with the subheadings

I. Strings for Strong Interactions

II. Non-string Kaluza-Klein Theories

III. Superstring Theories

I. Strings for Strong Interactions

This covers the period 1968-75 and will amount to a synopsis of an ancient book[1] published in 1974, out-of-print in 1979, and to be reprinted some time soon. The history starts with the Veneziano[2] model for the scattering of two particles into two particles (Fig. 1). The amplitude A(s, t) = A(t, s) is a crossing-symmetric function of the two Mandelstam invariants s, t. Veneziano actually chose the convenient process $\pi^a\pi^b \to \pi^c\omega$ with unique spin structure $\varepsilon_{abc}\varepsilon_{k\lambda\mu\nu}p_1^k p_2^\lambda p_3^\mu A(s, t)$. The question is what is A(s, t) as a function of s and t? This could be answered by a quantum field theory but in 1968 no such theory existed. The guess for A(s, t) was based on four phenomenological clues:

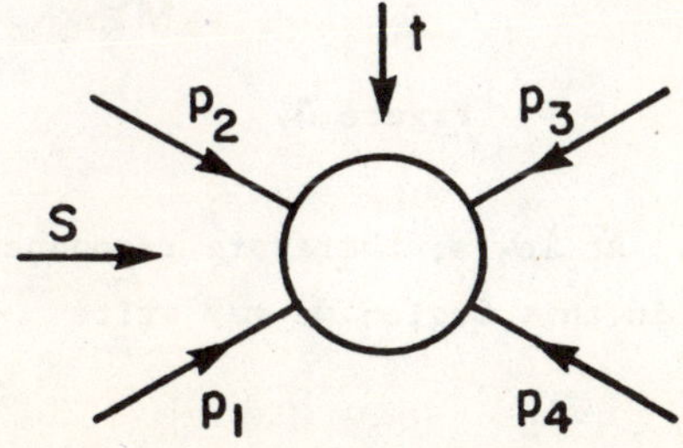

Figure 1.

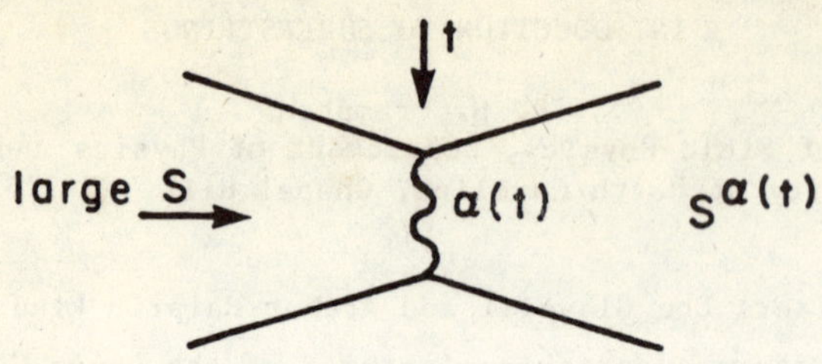

Figure 2.

(i) Regge behavior. For $s \to \infty$ with t fixed and spacelike (Fig. 2) we have

$$A(s, t) \sim s^{\alpha(t)} \tag{1}$$

where the Regge trajectory $\alpha(t) \simeq \alpha(o) + \alpha' t$ is approximately linear with Regge slope $\alpha' \simeq 0.9 \text{ GeV}^{-2}$. The mesons lie quite accurately on straight-line trajectories (e.g. Fig. 3.)

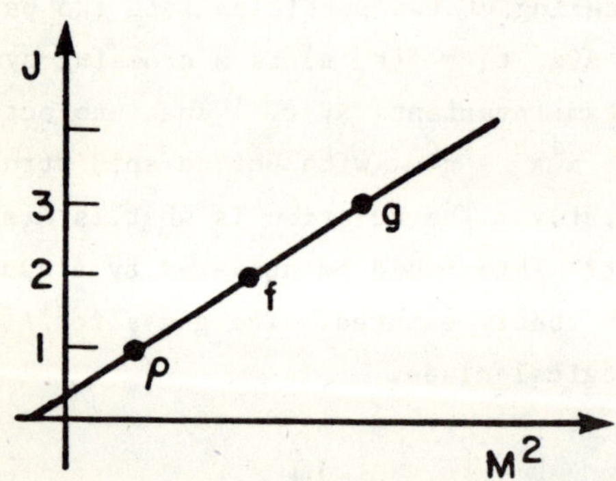

Figure 3.

(ii) Resonances. At low s, there are resonances with width << mass (Fig. 4) so that in this region we may write

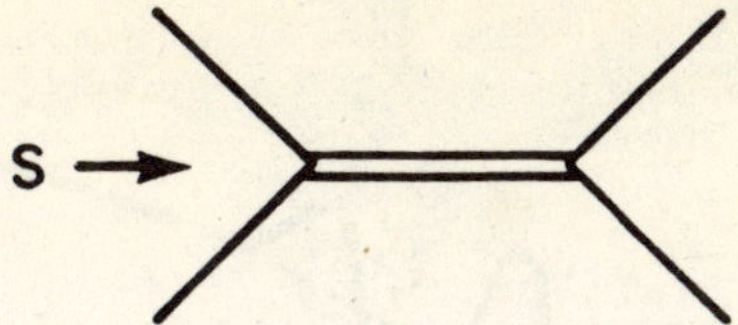

Figure 4.

$$A(s, t) = \sum_i \frac{a_i(t)}{s - s_i} \qquad (2)$$

with residues $a_i(t)$ polynomial in t of order equal to the resonance spin.

(iii) Analyticity. Because of Regge behavior and dispersion relations (analyticity), the absorptive part of A(s, t) satisfies[3] the Finite Energy Sum Rule (FESR)

$$\int^{\bar{s}} ds \, \mathrm{Im}A(s, t) = \left(\frac{\bar{s}}{s_o}\right)^{\alpha(t)+1} \frac{\beta(t)}{\alpha(t) + 1} \qquad (3)$$

Varying $\bar{s}$, the left hand side varies as $\bar{s}^{-1}$ (resonances) while the right side goes as $\bar{s}^{\alpha(t)+1}$. How is this possible while keeping A(s, t) = A(t, s)? was one conundrum before Veneziano. The relation of analyticity to linearly-rising Regge trajectories in a zero-width resonance approximation was discussed.[4]

(iv) Duality. A phenomenological question is: should one add the low-energy resonances to the high-energy Regge poles? Looking at the pion-nucleon charge exchange forward amplitude A(s, o), Dolen, Horn and Schmid showed that, phenomenologically, the Regge pole contribution, extrapolated back to low energy, gives an average description of the low-energy resonances (Fig. 5). Thus, addition of Regge plus resonances would be double-counting and instead there is a "duality" of the form

$$\langle \text{Resonances} \rangle_{\text{average}} = \text{Regge} \qquad (4)$$

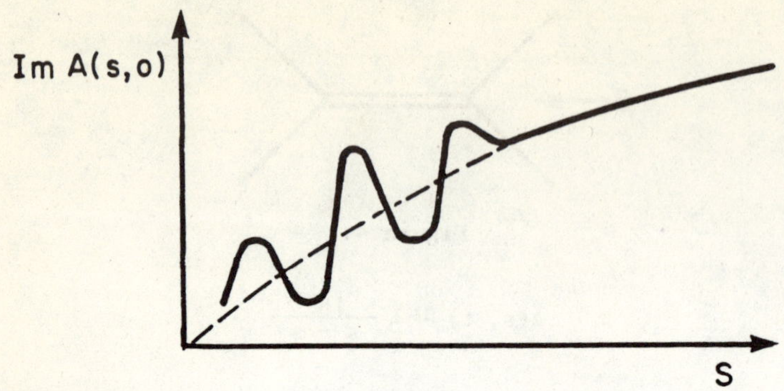

Figure 5.

One representation of this is through the quark diagrams of Harari[5] and Rosner[6], e.g. for meson-meson scattering (Fig. 6).

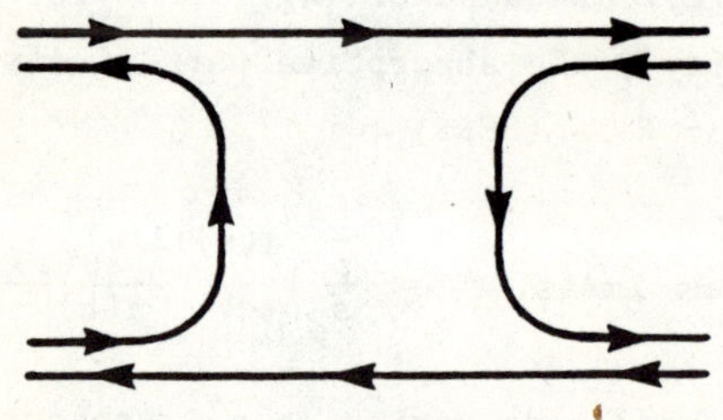

Figure 6

In elastic scattering there is a diffractive component leading to approximately constant total cross sections, through the optical theorem, and this is associated with the Pomeron singularity in the angular momentum plane. Just as the normal meson Regge trajectories are dual to resonances, the Pomeron appears phenomenologically[7] to be dual to non-resonant background, the relevant quark diagram being Fig. 7.

Based on these four clues Veneziano[2] made the guess that

$$A(s, t) = B\left(-\alpha(s), -\alpha(t)\right) \tag{5}$$

where

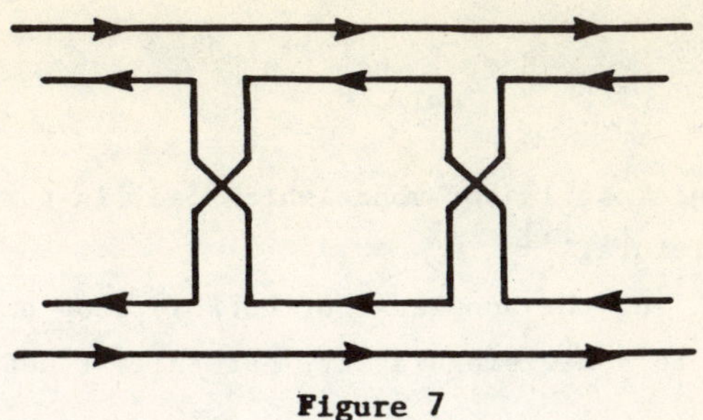

Figure 7

$$B(X, Y) = \frac{\Gamma(X)\Gamma(Y)}{\Gamma(X + Y)} \tag{6}$$

is the Euler B-function and $\alpha(s)$ is a linear function $\alpha(s) = \alpha(o) + \alpha' s$. This model has remarkable properties: manifest crossing-symmetry, expandable completely into poles in s (meromorphic), Regge behavior in all complex directions except the positive real axis, analyticity and duality. The question then was: what is the physics underlying Eq. (5)?

The first step was the generalization[8] to N-particle amplitudes (Fig. 8) incorporating similar duality properties. These amplitudes

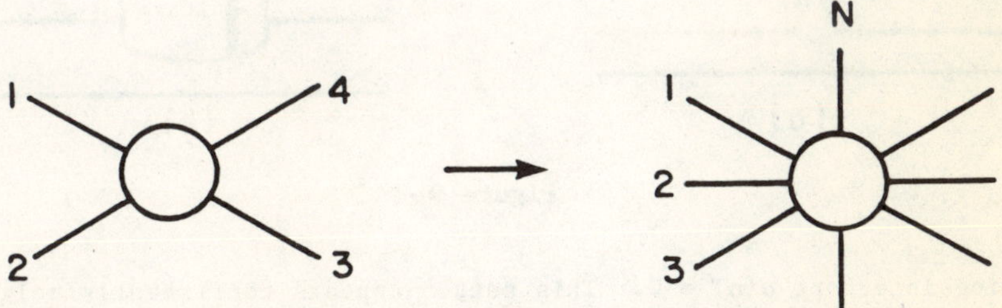

Figure 8

were then factorized[9] on a finite spectrum with degeneracy growing as $\sim e^{am}$, where m = mass, a = numerical constant. This then led to a covenient operator formalism[10] with Hamiltonian

$$H = \sum_{n=1}^{\infty} n a_\mu^{(n)+} a_\mu^{(n)} \tag{7}$$

with simple harmonic oscillator modes which led, in turn, to a physical string picture.[11]

Calculation[12] of the nonplanar orientable loop amplitude (Fig. 7) remarkably led to a new singularity, naturally (then) identified with the Pomeron.

The algebra of the 4-vector modes $(g_{\mu\nu} = \text{diag.}(+ ---))$

$$[a_\mu^{(m)}, a_\nu^{(u)+}] = - \delta_{mn} g_{\mu\nu} \tag{8}$$

means that the timelike excitations are ghosts. For spacetime dimension $d \leqslant 26$, however, all such ghosts decouple.[13] Examination of the Pomeron diagram which is redrawn in two different ways in Fig. 9 reveals a spectrum of closed string states with slope $\frac{1}{2}\alpha'$ and

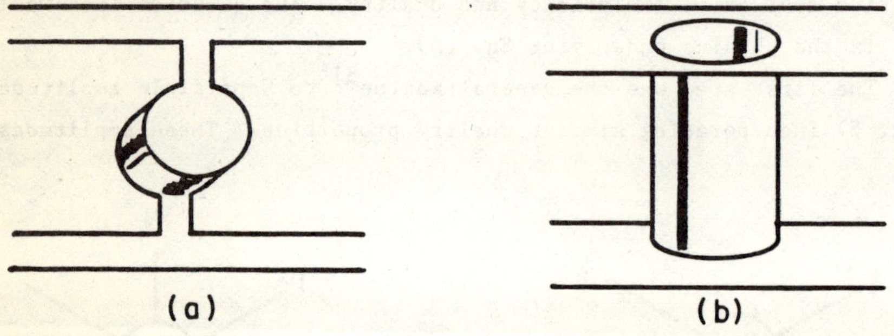

Figure 9

leading intercept $\alpha(o) = 2$. This sector appears consistently only for $d = 26$ exactly. When we scatter such closed strings we find a new type of singularity structure (Fig. 10) with a four-point function of the form[14]

$$A(s, t, u) = \frac{\Gamma(-\alpha_s/2)\Gamma(-\alpha_t/2)\Gamma(-\alpha_u/2)}{\Gamma(\frac{-\alpha_s-\alpha_t}{2})\Gamma(\frac{-\alpha_t-\alpha_u}{2})\Gamma(\frac{-\alpha_u-\alpha_s}{2})} \tag{9}$$

Figure 10

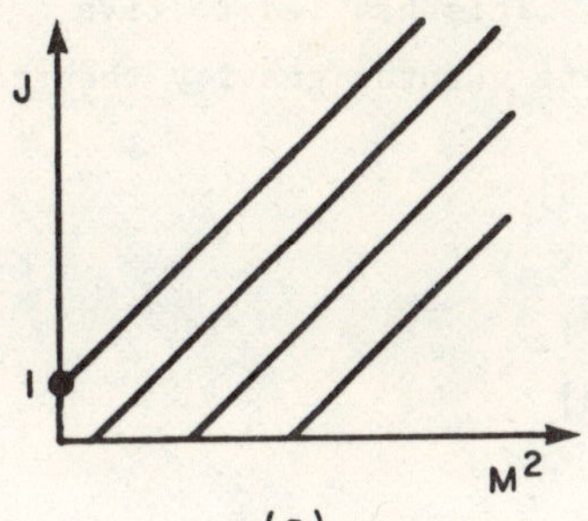

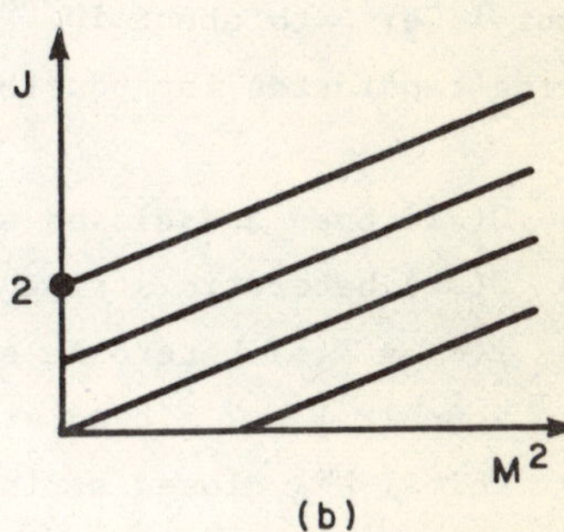

Figure 11

The spectra for open and closed strings are as indicated in Fig. 11. The intercepts required for strong interaction physics are $\alpha(o) \approx 1/2\ (\rho)$ and $\alpha(o) \approx 1$ (Pomeron), off by about a factor two from the dual resonance model prediction.

In 1971, a new dual model incorporating fermions was constructed[15] with the same intercepts but a critical spacetime dimension now $d = 10$. The odd G parity sector of this old model is precisely the O(32) superstring discussed in section III below.

Clearly the string does not provide a fundamental theory of strong interactions (alone). Aside from the serious difficulties of the spectrum and the dimension already given, the amplitudes for off-shell currents as well as the "granularity" needed for the parton model were other shortcomings. Finally, the success of QCD drove another nail into the coffin of the old string model which essentially died in 1975.

Nevertheless, we should bear in mind that hadrons do lie on linear Regge trajectories, there is phenomenological duality for hadron scattering, etc., so strings may approximate QCD in certain regimes, e.g., the linear quarkonium potential at large interquark separation.

The current rebirth of strings as applied to gravitational, and possibly all, interactions requires the identification[16] of the misplaced massless spin-two Pomeron with the graviton and decreases α' from about 1 GeV^{-2} to about 10^{-38} GeV^{-2}. This has led to five superstring candidates for possibly finite quantum gravity theories in d = 10:

(1) O(32) open and closed strings[15]

(2) O(32) heterotic string[17]

(3) E(8) x E(8) heterotic string[17]

(4) Nonchiral N=2 closed strings[18]

(5) Chiral N=2 closed strings[18]

But the phenomenological motivations from strong interactions have disappeared for quantum gravity. There are precedents, e.g. spontaneous symmetry breaking was used with some success in strong interactions before its crucial use in electroweak interactions. It may be that the old string people guessed a unique answer to the S-matrix problem and only now is it being used in the correct way.

II. Non-string Kaluza-Klein Theories

II.a. Anomalies

A principal difficulty of, and restriction on, all Kaluza-Klein theories is to obtain chiral fermions on M_4, four dimensional Minkowski space. This leads to two simple, but important, remarks:

(i) The initial dimension must be even d = 2n, with a spatial rotation group O(2n - 1), n = integer. In O(2n) rotational invariance implies that parity (P) is conserved because P is not a discrete operation but merely part of the rotation group e.g. for d = 3 and O(2) parity is a 180° rotation (Fig. 12). In such a case, no chirality can be defined.

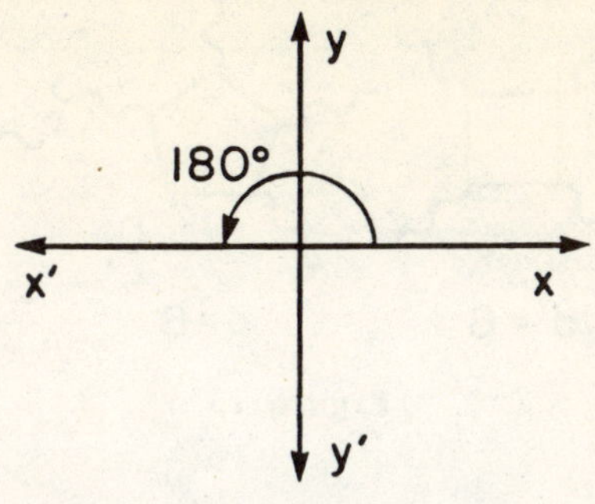

Figure 12

Equivalently, for d = 4, the four gamma matrices γ_μ can be combined to give the pseudoscalar γ^5 which allows the decomposition

$$\psi = \left(\frac{1 + \gamma_5}{2}\right)\psi + \left(\frac{1 - \gamma_5}{2}\right)\psi = \psi_R + \psi_L \tag{10}$$

In d = 5, by contrast, $\gamma^0, \gamma^1, \gamma^2, \gamma^3, \gamma^4$ form a complete set and there is no "γ^6".

(ii) There must be explicit gauge fields in the starting dimension d = 2n > 4. If we start with a chiral fermion ψ_L in d = 2n > 4 (eigenvalue of γ_{2n+1} equal to −1) then compactification to d = 4 with a compact manifold C_{2n-4} gives chiralities as follows

$$M_{2n} \rightarrow M_4 \times C_{2n-4}$$

$$\begin{aligned} \gamma_{2n+1} &= \gamma_5 \times \gamma_{2n-3} \\ &= + \;\; \times \;\; - \\ & - \;\; \times \;\; + \end{aligned} \tag{11}$$

so that both chiralities appear symmetrically on M_4. Thus the original Kaluza-Klein program starting with just higher dimensional gravity fails. One must introduce an asymmetry between the chiralities by arranging explicit gauge fields in a topologically nontrivial vacuum configuration on C_{2n-4}; this then allows survival of chirality on M_4.

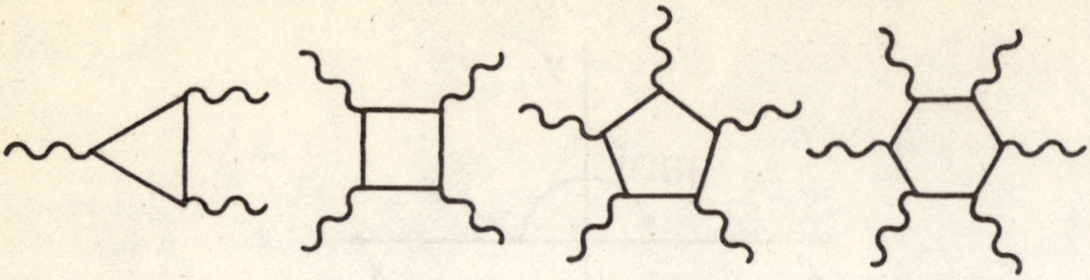

Figure 13

The necessity of explicit Yang-Mills fields in these theories was disappointing from the viewpoint of unification. One did, however, still hope to explain fermion quantum numbers, including the existence of families, by new anomaly cancellation conditions. This is because such gauge theories with chiral fermions have anomalies which must be zero in order to define the quantum theory and keep unitarity. The nonrenormalizability of such higher dimensional theories is irrelevant to the necessity for anomaly cancellation.

The leading contribution to the gauge anomaly in dimension $d = 2n$ is from a polygonal Feynman graph with $(n + 1)$ sides e.g. the triangle in $d = 4$ and the hexagon in $d = 10$ (Fig. 13). These anomalies may be computed using Fujikawa's method, or by differential geometry. We used[19] a brute force method of studying the linear divergence of the Feynman diagram (Fig. 14). Shifting integration variables appropriately one finds that the $5! = 120$ crossed diagrams combine in 60 pairs and enable one to compute

$$2q_k V_{k\lambda\mu\nu\rho\sigma}(p_a p_b p_c p_d p_e)$$

$$= 2^6 X_6 (ST)\, \varepsilon_{\lambda\mu\nu\rho\sigma\alpha\beta\gamma\delta\varepsilon} p_a^\alpha p_b^\beta p_c^\gamma p_d^\delta p_e^\varepsilon \tag{12}$$

with

$$X_6 = \frac{-1}{2^{10}\pi^5 5!} \tag{13}$$

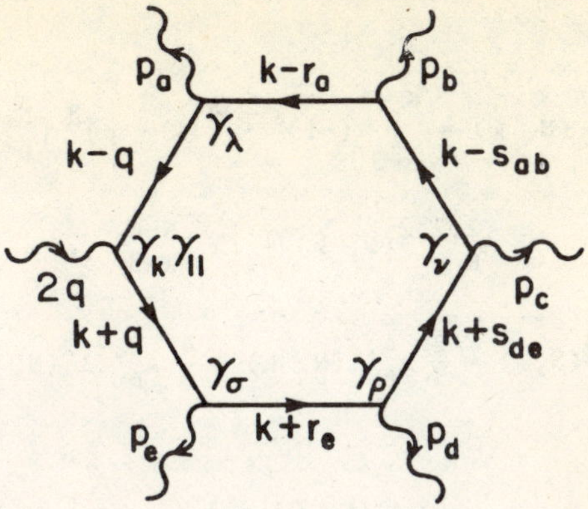

Figure 14

and

$$(ST) = STr(\Lambda^a\Lambda^b\Lambda^c\Lambda^d\Lambda^e) \tag{14}$$

is the symmetrized trace of generators in the fermion representation. We must decompose this trace into irreducible tensors formed from the defining representation

$$\begin{aligned} STr(\Lambda^a\Lambda^b\Lambda^c\Lambda^d\Lambda^e\Lambda^f) = {} & A_6 STr(\lambda^a\lambda^b\lambda^c\lambda^d\lambda^e\lambda^f) \\ & + A_6^{4,2} S\big(tr(\lambda^a\lambda^b\lambda^c\lambda^d)tr(\lambda^e\lambda^f)\big) \\ & + A_6^{3,3} S\big(tr(\lambda^a\lambda^b\lambda^c)tr(\lambda^d\lambda^e\lambda^f)\big) \\ & + A_6^{2,2,2} S(tr(\lambda^a\lambda^b)tr(\lambda^c\lambda^d)tr(\lambda^e\lambda^f)) \ . \end{aligned} \tag{15}$$

For SU(N) and k-rank antisymmetric tensors the relevant formulas for these hexagon anomalies are

$$A_6(N,k) = \sum_{p=0}^{k=1} (-1)^{k-p+1}(k-p)^5\binom{N}{p} \tag{16}$$

$$A_6^{4,2}(N,k) = 15\ A_4(N-2,\ k-1) \tag{17}$$

where

$$A_4(N, k) = \sum_{p=0}^{k=1} (-1)^{k-p+1} (k - p)^3 \binom{N}{p} \tag{18}$$

$$A_6^{2,2,2}(N, k) = 15 \binom{N-6}{k-3} \tag{19}$$

$$A_6^{3,3}(N,k) = \frac{2}{3} A_6^{4,2}(N, k) + \frac{4}{3} A_6^{2,2,2}(N, k) \tag{20}$$

For example:

$$A_6(N, 0) = 1 \tag{21}$$

$$A_6(N, 2) = (N - 32) \tag{22}$$

The k = 2 representation of SU(N) behaves like the adjoint of O(N). Only the disconnected (non-leading) anomalies may be compensated by Chern-Simons terms. O(32) is the <u>only</u> nonexceptional group for which the adjoint has leading hexagon anomaly $A_6 = 0$.

One immediate point is that if we are in d = 10 with only left-handed, and no right-handed, Weyl spinors then, without adding Chern-Simons classes, there cannot be anomaly cancellation for any nontrivial representation of any gauge group G. This can be seen by going into the Cartan subalgebra (diagonal) with real eigenvalues λ_i and noticing that

$$\sum_i \lambda_i^6 > 0 \tag{23}$$

The only possible cancellation in these nonstring cases is between left and right helicities (in d = 4k + 2). Thus, all supersymmetric Yang-Mills theories in d = 10 (or any d = 4k + 2) are anomalous. Because of the zero-slope limit it appeared that open superstrings were anomalous although we did say[20] in a footnote: "The only possible loophole would be if the infinite sum over massive fermions gives a nonvanishing anomaly precisely cancelling the anomaly of the massless fermions. We regard this as extremely unlikely." For O(32), however, this is precisely what happens from this viewpoint.

We now discuss those combinations of the representations [k] (K-rank antisymmetric) of SU(N) which have, in d = 2n, no gauge, gravity, or mixed chiral anomalies. For d = 6, the first non-trivial solution[21] ("non-trivial" means able to give chirality in d = 4 after compactification) is for SU(6) where there is only one possibility. For higher SU(N) there is one solution for SU(7), two for SU(8,9), three for SU(10,11), and so on. In general, for d = 2n and SU(N) with only antisymmetric representations the number of such independent solutions is

$$[\tfrac{1}{2}(N - n - 1)] \tag{24}$$

where [x] is the integer part of x.

There is a systematic procedure[22] for constructing these solutions: consider the superalgebra SU(N/M) with $(N - M) > (n + 1)$. Choose first a combination of $[k]^{(N-M)}$ which is real (n = even) or pure imaginary (n = odd). Now promote these Young tableaus $[k]^{(N-M)}$ to super Young tableaus $[k]^{(N/M)}$ defined by

$$[k]^{(N/M)} = \sum_{p=0}^{m} (-1)^p\left([k-p]^{(N)}, \{p\}^{(M)}\right) \tag{25}$$

where $\{p\}^{(M)}$ is a symmetric p-rank representation of SU(M) and the sign $(-1)^p$ determines the helicity of the state. The combination arrived at has no chiral anomalies of gauge or gravitational type.

It is convenient to introduce the principal representations[23,24]

$$P_{(2n)}(N, k) = [k]^{(N/N-2k+1)} + (-1)^n [k-1]^{(N/N-2k+1)} \tag{26}$$

and by taking

$$[\tfrac{1}{2}(n + 3)] < k < k_{max} = \begin{cases} [N/2] & n = \text{odd} \\ [{}^1\!/_2\,(N-1)] & n = \text{even} \end{cases} \tag{27}$$

we obtain the full number $[\frac{1}{2}(N - n - 1)]$ of independent solutions as conjectured by others. Note, in particular, that the number of independent choices is small.

II. b. Compactification

We seek solutions of the Einstein-Yang-Mills equations in $d = 2n$ with less than the maximal symmetry (M_{2n}). Actually, we shall discuss the manifolds of the form $M_4 \times S_{2n-4}$ (sphere).

Such classical solutions require that the energy-momentum tensor T_{MN} receive contributions from a nonzero vacuum value of the gauge field e.g. a monopole configuration for $n = 3$ (S_2), an instanton configuration for $n = 4$ (S_4), etc. This drives the compactification. At the same time, and quite attractively, this precisely enables survival of chiral fermions on M_4.

We shall require <u>classical</u> stability[25] of the vacuum under small perturbations--surely a necessary condition for stability in the complete quantum theory.

Consider first the case $d = 6$ with manifold $M_4 \times S_2$ and monopole-induced compactification.[26] The solution sought is therefore of the form

$$g_{MN} dz^M dz^N = g_{mn}(x) dx^m\, dx^n \pm g_{\mu\nu}(y) dy^\mu\, dy^\nu \tag{28}$$

$$g_{\mu\nu}(y) dy^\mu\, dy^\nu = a^2 (d\theta^2 + \sin^2\theta\, d\phi^2) \tag{29}$$

$$A_M dz^M = A_\mu(y) dy^\mu \tag{30}$$

$$= \frac{n}{2g} (\cos\theta \mp 1) d\phi \qquad \begin{matrix} 0 \leqslant \theta < \pi \\ 0 < \theta \leqslant \pi \end{matrix} \tag{31}$$

Here, n = monopole charge. It is worth pointing out that the monopole is not in the space: the field on S_2 is <u>as if</u> there were a monopole at its center, see Eq. (31).

To investigate which fields remain massless on M_4, we expand in monopole harmonics $Y_{q,LM}(\theta,\phi)$ which are eigenfunctions of $\underline{J}^2 = (\underline{L}^2 + q^2)$ with $q = ny/2$ where y is the U(1) charge of the field considered. For example, a massless (in $d = 6$) scalar field is expanded as

$$\Phi(x, y) = \sum_{JM} \phi_{JM}(x) Y_{q,JM}(\theta, \phi) \tag{32}$$

The only massless component (recall $J \geq |q|$) is for $q = 0$. If $n \neq 0$, this means $y = 0$. Recall that

$$\Box_z = \Box_x + \Box_y \tag{33}$$

so the eigenvalue of $\Box_y$ must be zero (s-wave, independent of θ and ϕ).

More interesting is a 4-component Weyl spinor Ψ_L in $d = 6$ with its expansion

$$\begin{pmatrix} \Psi_{L1} \\ \Psi_{L2} \\ \Psi_{L3} \\ \Psi_{L4} \end{pmatrix} = \sum_{JM} \begin{pmatrix} \Psi^1_{L,JM}(x)\, Y_{q - 1/2, JM}(\theta, \phi) \\ \Psi^2_{L,JM}(x)\, Y_{-q + 1/2, JM}(\theta, \phi) \\ \Psi^1_{R,JM}(x)\, Y_{q + 1/2, JM}(\theta, \phi) \\ \Psi^2_{R,JM}(x)\, Y_{-q - 1/2, JM}(\theta, \phi) \end{pmatrix} \tag{34}$$

For a given q, consider $J = |q| - 1/2$. Only the upper two ($q > 0$) or lower two ($q < 0$) components are nonzero. The corresponding multiplicity is $(2J + 1) = 2|q| = |ny|$. Hence we obtain $|ny|$ Weyl spinors massless on M_4 with the same ($q > 0$) or opposite ($q < 0$) helicity compared to the helicity in $d = 6$.

Using this rule, we can easily analyze the anomaly-free combinations.

It turns out that these $d = 6$ examples are not very successful phenomenologically ($d = 6$ supergravity may work better?) so I will go immediately to instanton-induced compactification in $d = 8$ with manifold $M_4 \times S_4$. This is similar to the $d = 6$ case except that now we write[27]

$$A^a_\mu(y) dy^\mu = \frac{2}{g} \frac{\eta_{a\mu\nu} y^2}{y^2 + b^2} dy^\mu \tag{35}$$

where μ = 5, 6, 7, 8 and a = 1, 2, 3 for the generators of SU(2)g. The full gauge group is SU(N) $\supset$ SU(2)g. Again we take Weyl spinors in antisymmetric representations $[k]^{(N)}$ of SU(N). These contain only t = 0 and t = 1/2 representations of SU(2)g. Writing

$$S_4 = \frac{SO(5)}{SO(4)} = \frac{SO(5)}{SU(2)_A \times SU(2)_B} \tag{36}$$

the invariance group of the instanton is $SU(2)_{B+g}$ and we must analyze an 8-component chiral spinor Ψ_L in d = 8 by expansion into irreducible representations of $SO(5) \supset SU(2)_A \times SU(2)_{B+g}$. We find, by a generalization of the procedure done explicitly above for d = 6, that

$$[k]_L^{(N)} \rightarrow \underset{SO(5)\ \text{singlet}}{[k-1]_L^{(N-2)}} \tag{37}$$

This is a very simple rule. In fact, it extends to complete superalgebra tableaus

$$[k]_L^{(N/M)} \rightarrow [k-1]^{(n-2/M)} \tag{38}$$

Now we ask: can we find completely anomaly-free fermions in d = 8 that give interesting chiral fermions on M_4? This is not a trivial question since the anomaly conditions are so restrictive. Nevertheless, the answer is affirmative.[23,24]

To arrive at SU(N) in d = 4 we start from SU(N + 2) in d = 8. Examples are:

$$N = 11 \qquad [5]_L^{(13/3)} + 2[4]_L^{(13/5)} + 3[3]_L^{(13/7)}$$

$$\rightarrow (11^4 + \overline{11}^3 + \overline{11}^2 + \overline{11}) \text{ of } SU(11) \tag{39}$$

which is a three-family model.[28]

$$N = 9 \qquad [4]_L^{(11/3)} + 3[3]_L^{(11/5)}$$

$$\rightarrow 9^3 + 9(\bar{9}) \text{ of } SU(9). \tag{40}$$

This also has three families[29] since on reduction to SU(5)

$$SU(9) \to SU(5)$$

$$[3]^{(9)} \quad : \quad 84 = \overline{10} + 4(10) + 6(5) + 4(1) \tag{41}$$

$$9[8]^{(9)} \quad : \quad 9(\bar{9}) = 9\left(\bar{5} + 4(1)\right) \tag{42}$$

so that adding, and dropping real representations that are expected to develop superheavy Dirac masses, leaves

$$3(10 + \bar{5}) \text{ of } SU(5). \tag{43}$$

The SO(5) isometry group is passive: all the massless chiral fermions on M_4 are SO(5) singlets. SO(5) nonsinglets are superheavy, near the Planck scale.

Finally, concerning nonstring Kaluza-Klein theory we should remark that the quantum fluctuations are expected to be large (the theory is non-renormalizable) and the only features of the classical solutions expected to survive are those massless particles protected by an unbroken symmetry like gauge vectors and chiral fermions. We should not take seriously any massless scalars which, since unprotected here by supersymmetry, will become superheavy due to quantum corrections.

Rather, we have shown how to obtain phenomenologically an acceptable mass spectrum for spin 1 and spin 1/2 in the effective theory on M_4.

Two objections to this approach are

1) the theory is non-unified; the gauge coupling is independent of the gravitational coupling.

2) the quantum corrections are out of control.

Both objections may be overcome by superstrings.

III. Superstring Theories

III.a. Anomaly Cancellation

Let us first focus on the anomaly cancellation[30] which occurs with the gauge group O(32). The d = 10 chiral fermions are in the 496 dimensional adjoint representation of O(32).

For the adjoint of O(N), put k = 2 in the SU(N) formulae already given to find

$$A_6(N, 2) = N - 32 \tag{44}$$

$$A_6^{4,2}(N, 2) = 15 \tag{45}$$

$$A_6^{2,2,2}(N, 2) = 0 \tag{46}$$

$$A_6^{3,3}(N,2) = 10 \tag{47}$$

But now we must realize that $tr(\lambda\lambda\lambda) = 0$ in O(N) if N ≠ 6 and hence we have

$$STr(\lambda\lambda\lambda\lambda\lambda\lambda) = (N - 32)Str(\lambda\lambda\lambda\lambda\lambda\lambda) + 15S\big(tr(\lambda\lambda\lambda\lambda)tr(\lambda\lambda)\big) \tag{48}$$

for Λ of O(N) in the adjoint representation. We see that O(32) supersymmetric Yang-Mills theory is not enough.

The field theory limit of the superstring has the bosonic terms

$$S = \int d^{10}x \sqrt{g}\,\Big[-\frac{1}{2\kappa^2} R - \frac{1}{\kappa^2}\frac{1}{\phi^2}\,\partial_\mu\phi\partial_\mu\phi$$

$$- \frac{1}{4g^2\phi} F^2 - \frac{3\kappa^2}{2g^4}\frac{1}{\phi^2} H^2\Big] \tag{49}$$

where

$$F = dA + A^2 \tag{50}$$

$$H = dB - \omega^o_{3Y} - \omega^o_{3L} \tag{51}$$

Note that the ten dimensional spacetime dictates that the dimensions are $[g] \sim L^3$, $[\kappa] \sim L^4$ so

$$\frac{\kappa}{g^2} \sim \frac{1}{L^2} \sim \frac{1}{\alpha'} \sim T \tag{52}$$

where α' = Regge slope and T = string tension. In fact, g^2 is absorbable into $\langle\phi\rangle$ leaving no dimensionless parameters at all, just T.

The presence of the Yang-Mills Chern-Simons form ω^{o}_{3Y} in Eq. (51) was discovered in N = 1 supergravity plus N = 1 super-Yang-Mills in d = 10 by two different groups[31] in 1982. One has

$$\omega^{o}_{3Y} = tr(AF - \frac{1}{3}A^{3}) \tag{53}$$

$$d\omega^{o}_{3Y} = tr\ F^{2} \tag{54}$$

$$\omega^{o}_{3L} = tr(\omega R - \frac{1}{3}\omega^{3}) \tag{55}$$

$$d\omega^{o}_{3L} = tr\ R^{2} \tag{56}$$

where the last two will be important in the gravitational anomalies.

Although $B_{\mu\nu}$ is in the supergravity multiplet, and hence a "gauge singlet," it does transform under Yang-Mills gauge transformation as dictated by local supergravity. Under

$$\delta A = d\Lambda + [A, \Lambda] \tag{57}$$

where Λ = gauge function, then

$$\delta\omega^{o}_{3Y} = d\omega^{1}_{2Y} \tag{58}$$

Corresponding to

$$TrF^{6} = (N - 32)trF^{6} + 15\ trF^{4}trF^{2} \tag{59}$$

the consistent anomaly is

$$G_{1} \sim \int (\frac{1}{3}\omega^{1}_{2Y}trF^{4} + \frac{2}{3}\omega^{1}_{6Y}trF^{2}) \tag{60}$$

If we add to the effective action

$$S_{1} = \int (BtrF^{4} + \frac{2}{3}\omega^{o}_{3Y}\omega^{o}_{7Y}) \tag{61}$$

with $\delta B = (-\omega^{1}_{2Y} - \omega^{1}_{2L})$ we find that

$$\delta S_1 = \int \delta B\ \mathrm{tr}F^4 - \frac{2}{3}\,(d\omega^1_{2Y})\omega^o_{7Y} - \frac{2}{3}\,\omega^o_{3Y}(d\omega^1_{6Y})) \tag{62}$$

$$= \int (\delta B\ \mathrm{tr}F^4 + \frac{2}{3}\,\omega^1_{2Y}\mathrm{tr}F^4 - \frac{2}{3}\,\omega^1_{6Y}\mathrm{tr}F^2) \tag{63}$$

$$= -\,G_1 - \int \omega^1_{2L}\mathrm{tr}F^4 \tag{64}$$

The miracle continues for O(32) since one finds that one $(3/2)_L$ gravitino plus one $(1/2)_R$--the content of the supergravity multiplet--plus n $(1/2)_L$ gives a gravitational anomaly[32] of the form

$$\frac{n-496}{7560}\,\mathrm{tr}R^6 + [\frac{1}{8} + \frac{n-496}{5760}]\ \mathrm{tr}R^4\mathrm{tr}R^2$$

$$+ [\frac{1}{32} + \frac{n-496}{13824}]\ (\mathrm{tr}R^2)^3 \tag{65}$$

so we must have n = 496 to cancel the leading gravitational anomaly = dimension O(32). By adding

$$S_2 = -\int (\frac{1}{32}\,B(\mathrm{tr}R^2)^2 + \frac{1}{8}\,B\ \mathrm{tr}R^4 + \frac{1}{12}\,\omega^o_{3L}\omega^o_{7L}) \tag{66}$$

and using $\delta\omega^o_{3L} = -\,d\omega^1_{2L}$, $\delta\omega^o_{7L} = -\,d\omega^1_{6L}$ we find

$$\delta S_2 = -\,G_2 + \int (\frac{1}{32}\,\omega^1_{2Y}(\mathrm{tr}R^2)^2 + \frac{1}{8}\,\omega^1_{2Y}\mathrm{tr}R^4) \tag{67}$$

where

$$G_2 = -\int (\frac{1}{32}\,\omega^1_{2L}(\mathrm{tr}R^2)^2 + \frac{1}{24}\,\omega^1_{2L}\mathrm{tr}R^4 + \frac{1}{12}\,\omega^1_{6L}\mathrm{tr}R^2) \tag{68}$$

is the consistent anomaly corresponding to Eq. (65).

The mixed gauge-gravity anomalies require two further numerical coincidences, making a total of four, in order that O(32) be an anomaly-free superstring.

For O(32) the mixed anomaly is

$$-\,\mathrm{tr}R^2[\mathrm{tr}F^4 + \frac{\hat{A}}{8}\,(\mathrm{tr}F^2)^2] + \hat{B}\mathrm{tr}F^2[\frac{1}{8}\,\mathrm{tr}R^4 + \frac{5}{32}\,(\mathrm{tr}R)^2] \tag{69}$$

where $\hat{A} = \hat{B} = 1$. For general O(N) one would have in this formula

$$\hat{A} = \frac{24}{N - 8} \quad \text{and} \quad B = \frac{\hat{N} - 2}{32} \tag{70}$$

The corresponding consistent anomaly is

$$\begin{aligned} G_3 = \int \Big(& \tfrac{1}{3}\, \omega^1_{2L} \mathrm{tr} F^4 + \tfrac{2}{3}\, \omega^1_{6Y} \mathrm{tr} R^2 \\ & + \frac{\hat{A}}{24}\, \omega^1_{2L} (\mathrm{tr} F^2)^2 + \frac{\hat{A}}{12}\, \omega^1_{2Y} \mathrm{tr} R^2 \mathrm{tr} F^2 \\ & - \frac{\hat{B}}{24}\, \omega^1_{2Y} \mathrm{tr} R^4 - \frac{\hat{B}}{12}\, \omega^1_{6L} \mathrm{tr} F^2 \\ & - \frac{5\hat{B}}{96}\, \omega^1_{2Y} (\mathrm{tr} R^2) + \frac{5\hat{B}}{48}\, \omega^1_{2L} \mathrm{tr} R^2 \mathrm{tr} F^2 \Big) . \end{aligned} \tag{71}$$

We now add to the effective action

$$\begin{aligned} S_3 = \int \Big(& \frac{K}{8}\, B \mathrm{tr} F^2 \mathrm{tr} R^2 + \frac{L}{48}\, \overset{o}{\omega}_{3L} \overset{o}{\omega}_{3Y} \mathrm{tr} R^2 \\ & - \frac{M}{24}\, \overset{o}{\omega}_{3Y} \overset{o}{\omega}_{3L} \mathrm{tr} F^2 - \frac{2N}{3}\, \overset{o}{\omega}_{3L} \overset{o}{\omega}_{7Y} + \frac{P}{12}\, \overset{o}{\omega}_{3Y} \overset{o}{\omega}_{7Y} \Big) \end{aligned} \tag{72}$$

where K, L, M, N, P are initially at our disposal. Now we wish to arrange that

$$\delta S_3 = - G_3 - (\delta S_1 + G_1) - (\delta S_2 + G_2) \tag{73}$$

and by looking at coefficients of particular terms we find:

$$\omega^1_{2L} (\mathrm{tr} F^2)^2 : \qquad M = \hat{A} \tag{74}$$

$$\omega^1_{2Y} \mathrm{tr} F^2 \mathrm{tr} R^2 : \qquad M = 3 - 2\hat{A} \text{ hence } \hat{A} = 1 \tag{75}$$

$$\omega^1_{2L} \mathrm{tr} F^2 \mathrm{tr} R^2 : \qquad L = 6 - 5\hat{B} \tag{76}$$

$$\omega^1_{2Y} (\mathrm{tr} R^2)^2 : \qquad 2L = 5\hat{B} - 3 \text{ and hence } \hat{B} = 1. \tag{77}$$

Hence we soon find that $\hat{A} = \hat{B} = 1$ both requiring O(32) uniquely. Of

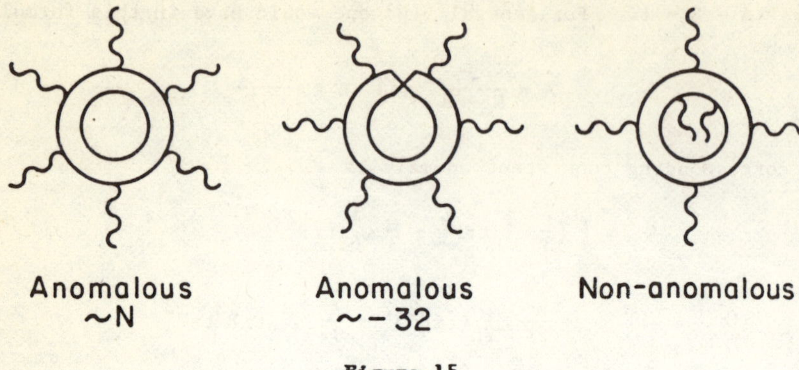

Figure 15

course, ones inevitable reaction is that one is discovering anomaly cancellations at too superficial a level and that all this must come from automatic anomaly-freedom in a higher-dimensional spacetime.

A similar series of anomaly cancellations works for the group E(8) x E(8) and the student is encouraged to work this out in similar detail as an exercise.

In 0(32), consider the amplitude with six external 0(32) massless gauge bosons. The three contributions are depicted in Fig. 15. The nonplanar orientable diagram (the final figure of Fig. 15) is non-anomalous because the hexagon anomaly corresponding to the first figure in Fig. 16 is cancelled by the massless tensor $(B_{\mu\nu})$ exchange shown also in Fig. 16. The $B_{\mu\nu}$ is in the same supermultiplet as the graviton and is a massless closed-string state. This diagram is really a "tree" rather than a "one-loop" diagram. The appearance of Planck's constant is complicated by the fact that closed strings have an intrinsic "h". Hence, the prejudice from quantum field theory that quantum corrections are loop effects is not true for the open super string loop expansion.

In any case, one has arrived at an anti-Kaluza-Klein theory of precisely the interesting type in terms of producing chiral fermions on M_4. Also, it is fully unified with gravity and may be a finite quantum theory (see below).

One may pause and ask how this new type of anomaly cancellation

Figure 16

impacts on nonstring Kaluza-Klein theory, in case we ever become disenchanted by superstrings. The answer is[33] that the non-leading gauge anomalies, the non-leading pure gravitational anomalies, and all the mixed gauge-gravitational anomalies can always be cancelled (at least for d = 6, 8, 10) by adding antisymmetric tensor fields transforming as suitable Chern-Simons classes. So one need cancel only the leading (connected) gauge anomaly in nonstring theories. The leading gravitational anomaly may be cancelled by gauge singlets that do not effect the light fermion content on M_4.

III.b. One Loop Finiteness of Open O(32) Superstring Theory

We shall consider one loop diagrams with some number M of external massless ground states of the superstring. This will imply all one loop graphs are finite by factorization if we can do any M. We shall use light-cone gauge. For example, for the bosonic modes the action in light-cone variables is just

$$S = -\frac{1}{4\pi\alpha'}\int_0^\pi d\sigma \int d\tau\, \partial_\alpha X^i \partial^\alpha X^i \tag{78}$$

The corresponding equations of motion are

$$\left(\frac{\partial^2}{\partial\sigma^2} - \frac{\partial^2}{\partial\tau}\right) X^i(\sigma, \tau) = 0 \tag{79}$$

with boundary conditions for an open string

$$\frac{\partial}{\partial\sigma} X^i = 0 \quad \text{at } \sigma = 0 \text{ and } \sigma = \pi \;. \tag{80}$$

The general solution is

$$X^i(\sigma, \tau) = x^i + p^i\tau + i \sum_{n\neq 0} \frac{1}{n} \alpha_n^i \cos n\sigma\, e^{-in\tau} \tag{81}$$

Upon quantization, one imposes

$$[\alpha_m^i, \alpha_n^j] = m\delta_{m+n,o}\delta^{ij} \tag{82}$$

Or defining, for $n > 1$, $\alpha_n^i = \sqrt{n}\, a_n^i$ and $\alpha_{-n}^i = \sqrt{n}\, a_n^{i+}$

$$[a_m^i, a_n^{j+}] = \delta_{mn}\delta^{ij} \tag{83}$$

Constraints give a mass-shell condition

$$\alpha'(\text{mass})^2 = N - 1 \tag{84}$$

$$N = \sum_{n=1}^{\infty} \alpha_{-n}^i \alpha_n^i = \sum_{n=1}^{\infty} n a_n^{i+} a_n^i \tag{85}$$

For superstrings, the $X^i(\sigma, \tau)$ are joined by spinor variables S^a which satisfy the subsidiary conditions in light-cone gauge

$$(h^-)^{ab}S^b = (\gamma^+)^{ab}S^b = 0 \tag{86}$$

where $h^- = \frac{1}{2}(1 - \gamma^{11})$. As discussed in Ref. 34, these conditions together with a Majorana condition and the Dirac equation reduce S^a to only <u>eight</u> real components. Ab initio there are 64 complex components--2-spinors on the 2-spacetime of the world sheet and 32-spinors on the 10-spacetime manifold; these 128 real components are halved four times by the Weyl condition, the Majorana condition, the light-cone (γ^+) constraint and finally the Dirac equation. At the endpoint $\sigma = 0$ we expand

$$S^a = \sum_{n=-\infty}^{\infty} S_n^a e^{-in\tau} \tag{87}$$

and quantize by

$$\{S^a_m, \bar{S}^b_n\} = (\gamma^+ h^-)^{ab} \delta_{m+n,o} \tag{88}$$

$$[\alpha^i_m, S^a_n] = 0 \,. \tag{89}$$

The superstring mass shell condition is

$$\alpha'(\text{mass})^2 = N = \sum_{n=1}^{\infty} (n a^{i\dagger}_n a^i_n + \frac{n}{2} \bar{S}_{-n} \gamma^- S_n) \tag{90}$$

The ground state of the open superstring is comprised of 16 massless states: 8 form a vector

$$|i\rangle \qquad 1 \leq i \leq 8 \tag{91}$$

$$\langle i\,|j\rangle = \delta_{ij} \tag{92}$$

Since $[S_o, N] = 0$ we may act with S_o to obtain a massless Majorana-Weyl spinor

$$|a\rangle = \frac{i}{8} (\gamma_i S_o)^a \,|i\rangle \quad 1 \leq a \leq 8 \tag{93}$$

The S^a_o satisfy

$$S^a_o \bar{S}^b_o = \frac{1}{2} (\gamma^+ h^-)^{ab} + \frac{1}{4} (\gamma^{ij+} h^-)^{ab} R^{ij}_o \tag{94}$$

where γ^{ij+} is the fully antisymmetrized produce of $\gamma^i \gamma^j \gamma^+$ and

$$R^{ij}_o = \frac{1}{8} \bar{S}_o \gamma^{ij-} S_o \tag{95}$$

are generators of O(8) in the transverse space.

The excited massive superstring states may be assembled by acting on the ground state with the creation operators α^i_{-n} and S^a_{-n} . For example, at N = 1, there are 128 boson states

$$\alpha^i_{-1}|j\rangle \text{ and } S^a_{-1}|b\rangle \tag{96}$$

and 128 fermion states

$$\alpha^i_{-1}|a\rangle \text{ and } S^a_{-1}|i\rangle \tag{97}$$

The student should verify there are 1152 of each at N = 2.

It is sometimes useful in computations to define

$$\phi_n^a = \frac{1}{2^{3/4}} \gamma^o \gamma^- S_n^a = \frac{1}{2^{1/4}} S_n^a \tag{98}$$

satisfying the simple anticommutation relations

$$\{\phi_n^a, \phi_n^{b+}\} = S^{ab} \tag{99}$$

Since

$$\frac{1}{2} \bar{S}_{-n} \gamma^- S_n = \phi_n^+ \phi_n \tag{100}$$

it follows that e.g.

$$\mathrm{tr}\, w^{1/2n\bar{S}_{-n}\gamma^- S_n} = \mathrm{tr}\, w^{n\phi_n^+\phi_n} = (1 - w^n)^8 \tag{101}$$

$$\mathrm{tr}\, w^{N_F} = \prod_{n=1}^{\infty} (1 - w^n)^8 \tag{102}$$

where

$$N_F = \sum_{n=1}^{\infty} \frac{1}{2} n\bar{S}_{-n} \gamma^- S_n \; . \tag{103}$$

In order to calculate our one-loop superstring graphs with external massless gauge bosons we shall need only the vertex for emitting such a state. In the bosonic string, the massless vector state is

$$|i\rangle = \alpha_{-1}^i |0\rangle \tag{104}$$

If $\zeta^\mu(k)$ is the polarization vector with $k \cdot k = k \cdot \zeta = 0$ then the appropriate vertex is

$$V = g\zeta \cdot P(\tau) e^{ik \cdot X(\tau)} \tag{105}$$

$$P^\mu(\tau) = \frac{\partial}{\partial \tau} X^\mu(\tau)\Big|_{\sigma=0} \tag{106}$$

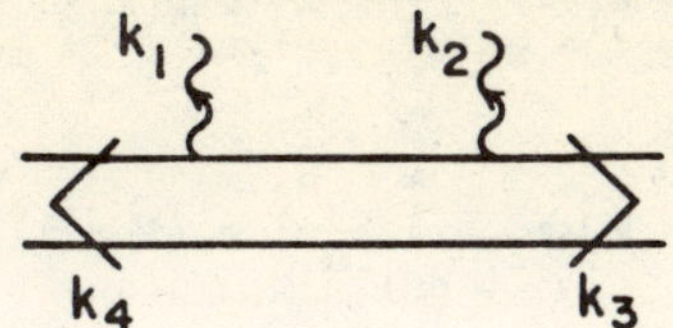

Figure 17

In light cone gauge, with $k^+ = 0$ and $\zeta^+ = 0$ this is just $g\zeta^i P^i V_o(k)$. For the superstring the relevant vertex is

$$g\zeta^i(P^i + k^j R^{ij})V_o(k) \tag{107}$$

where $R^{ij} = \frac{1}{8}\bar{S}\gamma^{ij-}S$. Here there is no ordering ambiguity because $k\cdot k = k\cdot\zeta = 0$. Although light-cone gauge itself is not restrictive, the fact that we are putting $k_+ = 0$ means that we may deal only with $M \leq 10$ external lines.

For a beginning, consider the bosonic string tree amplitudes with external spinless tachyon ground states. The vertex is

$$V_o(k) = g\colon e^{ik\cdot X(\tau)}\colon \tag{108}$$

$$= g\,\exp\Big(\sum_{n=1}^{\infty}\frac{k\cdot a_n^+}{\sqrt{n}}\Big)\exp\Big(-\sum_{n=1}^{\infty}\frac{k\cdot a_n}{\sqrt{n}}\Big) \tag{109}$$

The propagator is

$$\Delta = \frac{1}{2}(L-1)^{-1} = \int_0^1 dx\, x^{L-2} \tag{110}$$

$$L = N + \frac{1}{2}p^2 \tag{111}$$

$$N = \sum_{n=1}^{\infty}\alpha_{-n}^i\alpha_n^i = \sum_{n=1}^{\infty} n a_n^{i+}a_n^i \tag{112}$$

For the four-point function of Fig. 17 the prototypical string calculation is just a "vacuum" value; we regard the k_3 and k_4 tachyons as unexcited Fock space vacua and compute

$$\langle 0|\ \exp\left(-\, k_1\cdot \sum_{n=1}^{\infty} \frac{\alpha_n}{n}\right) x^N \exp\left(k_2\cdot \sum_{m=1}^{\infty} \frac{\alpha_{-m}}{m}\right)|0\rangle$$

$$\exp\left(-\, k_1\cdot k_2 \sum_{n=1}^{\infty} \frac{x^n}{n}\right) = (1-x)^{k_1\cdot k_2} \tag{113}$$

and we may write

$$A_4 = \int_0^1 dx\ x^{-\frac{1}{2}s-2}\ (1-x)^{-\frac{1}{2}t-2} \tag{114}$$

for which the crossing symmetry between s and t was the first "string miracle." The M point function is obtained by an extension of this, and is most conveniently written with M complex variables z_i on a unit circle (Fig. 18) in the form

$$A_M = \int \frac{dz_i}{dV_3} \prod_{1\leq I<J\leq M} (z_I - z_J)^{k_I\cdot k_j} \tag{115}$$

with

$$dV_3 = \frac{dz_a\, dz_b\, dz_c}{(z_a - z_b)(z_b - z_c)(z_c - z_a)} \tag{116}$$

with the z_a, z_b, z_c arbitrarily chosen, reflecting the 3-parameter Moebius invariance of an SO(2,1) group.

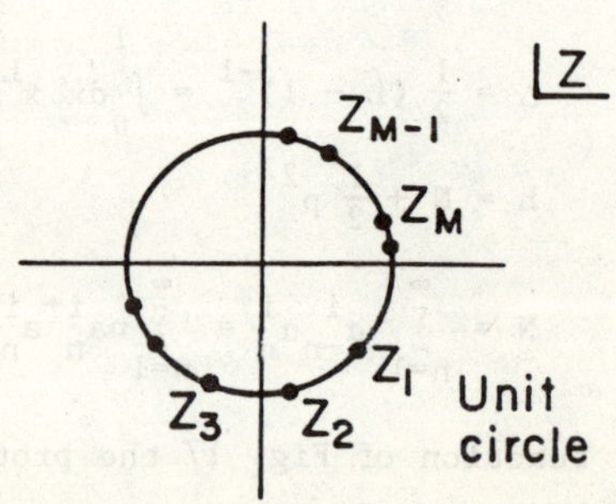

Figure 18

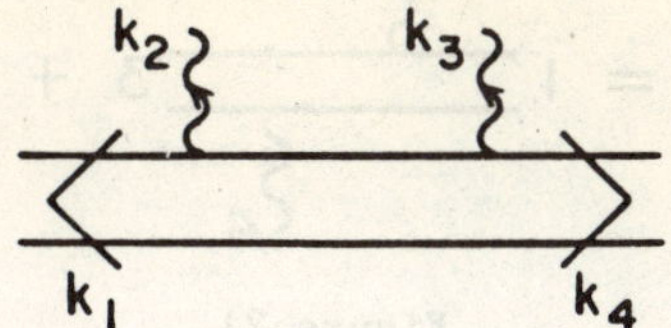

Figure 19

For the open superstring amplitude of Fig. 19 the S_o modes are very crucial. The relevant matrix element for four external massless gauge bosons is

$$\langle k_1, \zeta_1 | V(k_2, \zeta_2)\, \Delta\, V(k_3, \zeta_3) | k_4, \zeta_4 \rangle \tag{117}$$

with $V(k, \zeta)$ given by Eq. (107). The result is the crossing symmetric amplitude

$$A_4 = K_4 \frac{\Gamma(-\tfrac{1}{2}s)\Gamma(-\tfrac{1}{2}t)}{\Gamma(-\tfrac{1}{2}s - \tfrac{1}{2}t)} \tag{118}$$

with the kinematic factor given by

$$K_4 = \zeta_1^{i_1} \zeta_2^{i_2} \zeta_3^{i_3} \zeta_4^{i_4}\, k_1^{j_1} k_2^{j_2} k_3^{j_3} k_4^{j_4}\, \mathrm{tr}\left(R_o^{i_1 j_1} R_o^{i_2 j_2} R_o^{i_3 j_3} R_o^{i_4 j_4}\right) \tag{119}$$

Derivation of Eq. (118) is a good exercise.

We shall need the twisting operator in our discussion of the Moebius strip superstring loops. In the tree diagrams Fig. 20

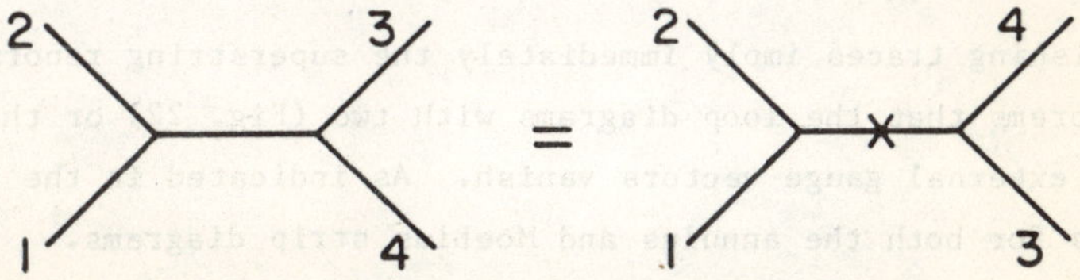

Figure 20

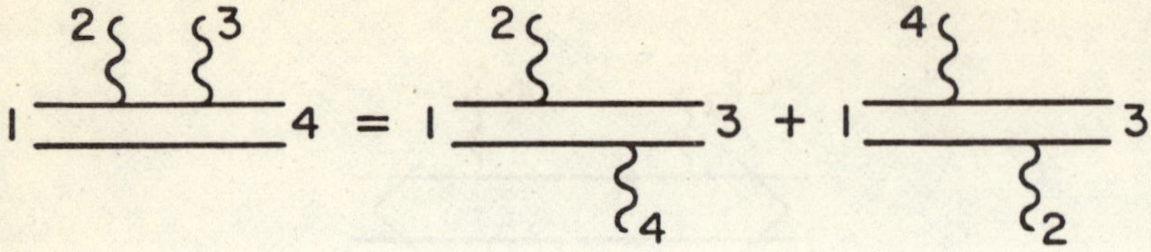

Figure 21

provides a simplistic notion. It has potential pit-falls; for example, if Ω is the twisting operator we have for the four-point tree amplitude

$$\langle 1|V(2)\Delta V(3)|4\rangle = \langle 1|V(2)\Delta\Omega V(4)|3\rangle$$
$$+ \langle 1|V(4)\Delta\Omega V(2)|3\rangle \tag{120}$$

as suggested in Fig. 21. The point here is that the full amplitude is given by Eq. (114) while the pieces on the right hand side of Eq. (120) represent

$$\int_0^1 = \int_0^{1/2} + \int_{1/2}^1 \tag{121}$$

In practice, we may evaluate only one twisted graph, then adjust the region of integration to account for the number of twisted graphs. This is an awkwardness of the operator formalism which can be rendered unique in path integrals.

The operators R_o^{ij} of Eqs. (94), (95) play an important role in superstring loop diagrams since one can show that in the S_o^a space

$$\mathrm{Tr}(1) = \mathrm{Tr}(R_o) = \mathrm{Tr}(R_oR_o) = \mathrm{Tr}(R_oR_oR_o) = 0. \tag{122}$$

These vanishing traces imply immediately the superstring renormalizatin theorems that the loop diagrams with two (Fig. 22) or three (Fig. 23) external gauge vectors vanish. As indicated in the Figures, this holds for both the annulus and Moebius strip diagrams.

On the other hand, since $\mathrm{tr}(R_oR_oR_oR_o)$ does <u>not</u> vanish, the diagrams with four external gauge bosons do not vanish (Fig. 24). Actually these diagrams diverge due to soft dilaton emission into the

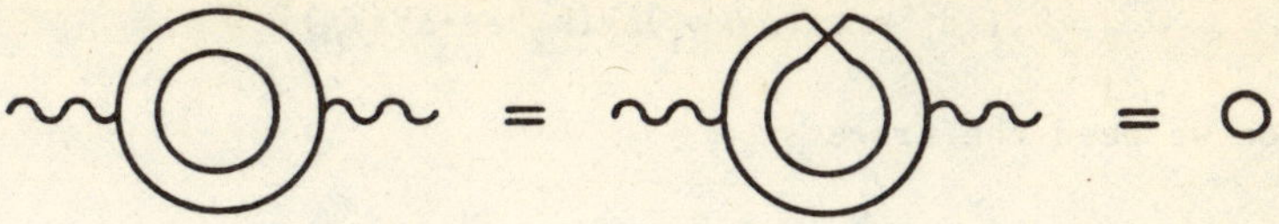

Figure 22

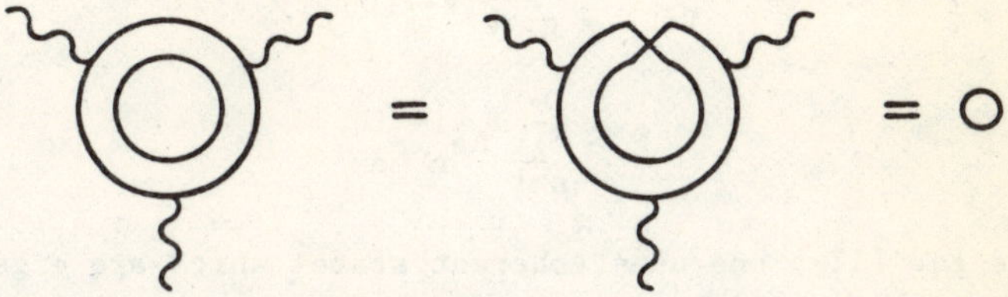

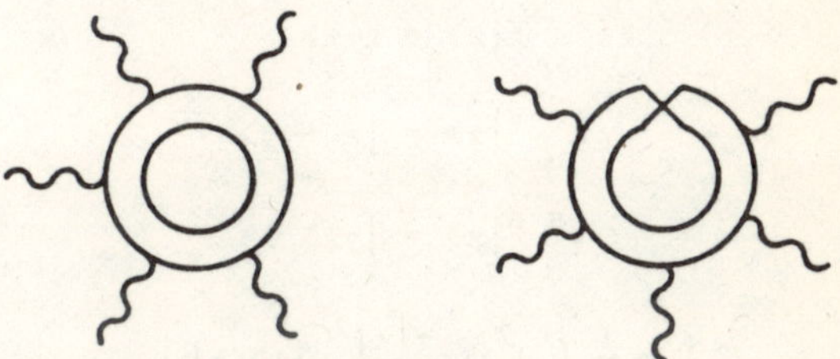

Figure 23

Figure 24

vacuum.

We are now ready to compute string loop diagrams. The technique is old,[35] dating from 1969, but unfamiliar to those of you who have computed one loop graphs only in quantum field theory. But once one has computed a one-loop graph for the bosonic-string, the essential stringiness of the loop has been done and one-loop superstring graphs are only slightly different. We shall spend the rest of these lectures discussing one-loop string graphs.

Consider the bosonic string in d = 26 with M external ground state tachyons emitted from the outer boundary of an annular string loop. We must compute the amplitude

$$\int d^{26}p \; \mathrm{tr}\left(\Delta V(k_1)\Delta V(k_2)\cdots\Delta V(k_M)\right) \tag{123}$$

and hence we need the trace

$$\mathrm{Tr}\left[x_1^N V(k_1) x_2^N V(k_2)\cdots x_M^N V(k_M)\right] \tag{124}$$

where

$$V(k) = g\!:e^{ik\cdot x(0)}\!: \tag{125}$$

$$N = \sum_{n=1}^{\infty} n a_n^{i+} a_n^{i} \tag{126}$$

To evaluate Eq. (124) one uses coherent states which are eigenstates of the annihilation operator. The basic formulas are

$$|z\rangle = e^{a^+ z}|0\rangle \tag{127}$$

$$a|z\rangle = z|z\rangle \tag{128}$$

$$e^{a^+ y}|z\rangle = |z + y\rangle \tag{129}$$

$$y^{a^+ a}|z\rangle = |yz\rangle \tag{130}$$

$$1 = \frac{1}{\pi}\int d^2\lambda \; e^{-|\lambda|^2} \; |\lambda\rangle\langle\lambda| \tag{131}$$

$$\mathrm{tr}A = \frac{1}{\pi}\int d^2\lambda \; e^{-|\lambda|^2} \; \langle\lambda|A|\lambda\rangle \tag{132}$$

where a, y, λ are complex numbers. Hence the trace of Eq. (124) can be written

$$g^M \prod_{n=1}^{\infty} \prod_{i=1}^{D-2} T_{ni} \tag{133}$$

where

$$T_{ni} = \int \frac{d^2\lambda}{\pi} e^{-|\lambda|^2} \langle\lambda| x_1^{na^+a} \exp(\frac{k_1^i a^+}{\sqrt{n}}) \exp(-\frac{k_1^i a}{\sqrt{n}}) x_2^{na^+a} \cdots$$

$$\cdots x_M^{na^+a} \exp(\frac{k_M^i a^+}{\sqrt{n}}) \exp(-\frac{k_M^i a}{\sqrt{n}}) |\lambda\rangle \tag{134}$$

where $a = a_n^i$. Now define $q_I = k_I/\sqrt{n}$, $u_I = x_I^n$ and

$$O_I = u_I^{a^+a} e^{q_I a^+} e^{-q_I a} \tag{135}$$

then

$$O_I|z\rangle = e^{-q_I z} |u_I(z + q_I)\rangle \tag{136}$$

and

$$O_L O_{L+1} \cdots O_M |z\rangle = e^{-\beta_L z} |z_L\rangle \tag{137}$$

defines Z_L and β_L. These satisfy the recursion relations

$$z_L = u_L(z_{L+1} + q_L) \tag{138}$$

$$\beta_L = q_L z_{L+1} + \beta_{L+1} \tag{139}$$

In particular, defining

$$z_1 = Wz + C \tag{140}$$

$$\beta_1 = - Bz - D \tag{141}$$

the required trace is

$$\frac{1}{\pi} \int d^2z\, e^{-|z|^2} e^{-\beta_1} \langle z|z_1\rangle =$$

$$= \frac{1}{\pi} \int d^2z \exp(-|z|^2 - \beta_1 + z^* z_1) \tag{142}$$

$$= \frac{1}{\pi} \int d^2z \exp[(1 - W)|z|^2 + Bz + Cz^* + D] . \tag{143}$$

Putting $z = x + iy$ the Gaussian integrals result in

$$\frac{1}{1-W}\exp\left(\frac{-D(W-1)+BC}{1-W}\right)$$

$$=\frac{1}{1-W}\exp\left(\frac{-\sum_{I,J=1}^{M} q_I^i q_J^i C_{IJ}}{1-W}\right) \tag{144}$$

where

$$C_{IJ} = \begin{matrix} (u_1u_2\cdots u_I)(u_{J+1}\cdots\cdots u_M) & (I < J) \\ u_{J+1}u_{j+2}\cdots u_I & (I > J) \end{matrix} \tag{145}$$

Here we used the solutions of Eqs. (138) and (139), namely

$$W = \prod_{I=1}^{M} u_I \tag{146}$$

$$C = \sum_{I=1}^{M} q_I(u_1u_2\cdots u_I) \tag{147}$$

$$B = -\sum_{k=1}^{M} q_k \sum_{I=k+1}^{M+1} u_I \tag{148}$$

$$D = -\sum_{1<K<J<M} q_K q_J (u_{K+1}\cdots u_J) \tag{149}$$

with $u_{M+1} \equiv 1$. Taking the product over modes and dimensions our trace becomes

$$g^M(f(w))^{2-D} \prod_{n=1}^{\infty} \prod_{I<J} \exp\left(\frac{-k_I\cdot k_J(C_{JI}^n + (w/C_{JI})^n - 2w^n)}{n(1-w^n)}\right) \tag{150}$$

where $w = x_1x_2\cdots x_m$ and

$$f(w) = \prod_{p=1}^{\infty}(1-w^p)\ . \tag{151}$$

We may further massage Eq. (150) by doing the logarithmic sums over n inside the exponent giving

$$g^M f(w)^{2-D} \prod_{I<J} \Big[\prod_{p=1}^{\infty} \frac{(1 - w^{p-1}C_{JI})(1 - w^p/C_{JI})}{(1 - w^p)^2} \Big]^{k_I \cdot k_J} \tag{152}$$

The momentum integral is

$$\int d^D p \prod_{i=1}^{M} x_I^{p_I^2/2} \tag{153}$$

where

$$p_I = p - \sum_{J=1}^{I-1} k_J = p + \sum_{J=I}^{M} k_J \tag{154}$$

The integrand may be written

$$\exp\Big[\sum_{I=1}^{M} \ln x_I \big(\tfrac{1}{2} p^2 + p \sum_{J=1}^{M} k_J\big) - \tfrac{1}{2} \sum_{I<J} k_I \cdot k_J \ln C_{JI} \Big] \tag{155}$$

and gaussian integration then gives, for Eq. (153),

$$\left(- \frac{2\pi}{\ln w}\right)^{D/2} \prod_{I<J} \Big[C_{IJ}^{-1/2} \exp\Big(\frac{\ln^2 C_{IJ}}{2\ln w}\Big) \Big]^{k_I \cdot k_J} \tag{156}$$

Incidentally, the gaussian integral we keep using is

$$\int_{-\infty}^{\infty} dx\, e^{-a^2x^2 \pm bx} = \frac{\sqrt{\pi}}{a} \exp\Big(\frac{b^2}{4a^2}\Big) \tag{157}$$

Collecting results, the bosonic annulus with M external spinless tachyons has amplitude

$$A_M = g^M \int \prod_{I=1}^{M} dx_I\, w^{-2} f(w)^{-24} \left(- \frac{2\pi}{\ln w}\right) \prod_{I<K} (\psi_{IJ})^{k_I \cdot k_J} \tag{158}$$

where

$$\psi_{IJ} = \psi(C_{JI}, w) \tag{159}$$

$$\psi(x, w) = \frac{1}{\sqrt{x}} \exp\Big(\frac{\ln^2 x}{2\ln w}\Big) \prod_{n=1}^{\infty} \frac{(1 - w^{n-1}x)(1 - w^n/x)}{(1 - w^n)^2} \tag{160}$$

It is now useful to transform to the disc variables

$$\nu_I = \frac{\ln(x_1 x_2 \cdots x_I)}{\ln w} \tag{161}$$

$$q = \exp\left(\frac{2\pi^2}{\ln w}\right) \tag{162}$$

so

$$d\nu_I = \frac{1}{\ln w} \sum_{J=1}^{I} \frac{dx_J}{x_J} + \frac{\ln(x_1 \cdots x_J)}{2\pi^2} \frac{dq}{q} \tag{163}$$

$$\frac{dq}{q} = -\frac{2\pi^2}{(\ln w)^2} \sum_{I=1}^{M} \frac{dx_I}{x_I} \tag{164}$$

giving

$$\sum_{I=1}^{M} dx_I = \frac{1}{2\pi^2} w(-\ln w)^{M+1} \frac{dq}{q} \prod_{I=1}^{M-1} d\nu_I \theta(\nu_{I+1} - \nu_J) \tag{165}$$

In these disc variables the amplitude looks like

$$A_M = g^M \int_0^1 \prod_{I=1}^{M-1} d\nu_J \; \theta(\nu_{I+1} - \nu_I)$$

$$\int \frac{dq}{q^3} \left(-\frac{2\pi^2}{\ln q}\right)^M f(q^2)^{-24} \prod_{I<J} (\psi_{IJ})^{k_I \cdot k_J} \tag{166}$$

As $q \to 0$, the hole at the center of the annulus shrinks to a point and one picks up the closed string pole corresponding to the massless scalar dilaton coupling to the vacuum. Using

$$\psi_{IJ} \sim (\ln q)^{-1} \tag{167}$$

$$\sum_{I<J} k_I \cdot k_J = + M \tag{168}$$

(since $k_I^2 = + 2$) we see the integrand of Eq. (166) is meromorphic at $q = 0$. Actually it is a higher order pole here because of tachyons; also, the $q^2 \to 1$ singularity of $f(q^2)^{-24}$ is severe.

Rather than continuing to discuss the unpleasant singularities of Eq. (166) we go straight to the open superstring. Consider the annulus with M = 4. Firstly we find a factor (the propagator is $(\frac{1}{2}p^2 + N_B + N_F)^{-1}$)

$$\mathrm{tr}\,(w^{N_F}) = f(w)^{+8} \tag{169}$$

which precisely cancels the $f(w)^{-8}$ from the bosonic modes. Hence supersymmetry removes the worst divergence immediately. Taking the gauge group O(N) the result is[36]

$$A_4^{(\mathrm{Annulus})} = NK_4 \int_0^1 \frac{dq}{q} \int_0^1 \prod_{I=1}^{3} d\nu_I\, \theta(\nu_{I+1} - \nu_I) \prod_{I<J} (\psi_{IJ})^{k_I \cdot k_J} \tag{170}$$

where N is the group factor coming from the empty boundary and

$$K_4 = 16\pi^3 g^4\, \mathrm{Tr}(\lambda^{a_1}\lambda^{a_2}\lambda^{a_3}\lambda^{a_4})\, t^{i_1j_1i_2j_2i_3j_3i_4j_4}$$

$$\zeta_1^{i_1}\zeta_2^{i_2}\zeta_3^{i_3}\zeta_4^{i_4}k_1^{j_1}k_2^{j_2}k_3^{j_3}k_4^{j_4} \tag{171}$$

$$t^{i_1j_1i_2j_2i_3j_3i_4j_4} = \mathrm{tr}(R_o^{i_1j_1}R_o^{i_2j_2}R_o^{i_3j_3}R_o^{i_4j_4}) \tag{172}$$

Note especially that the $(\ell nq)^{-M-1}$ from the Jacobian cancels the $(\ell nq)^5$ from the momentum integral $\int d^{10}p$. Also, since $\sum_{I<J} k_I \cdot k_J = 0$ for massless vector external states, the integrand is meromorphic at q = 0. It is now a simple dilaton pole since tachyons are also eliminated.

The M = 4 Moebius strip (Fig. 25) has no N since there is no empty boundary. The twisting operator on the propagator is

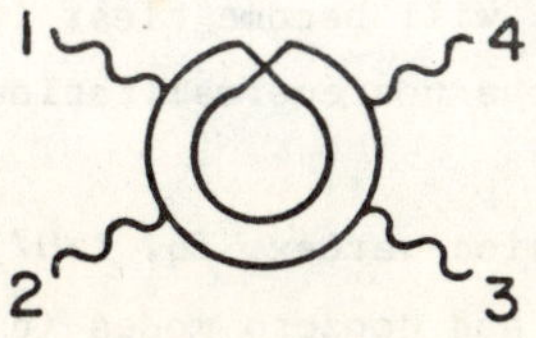

Figure 25

$$\Omega = (-1)^{N_B + N_F} \tag{173}$$

so it makes $x_1 \to -x_1$ and $w \to -w$. The integration range of ν_1 becomes (0, 2) instead of (0, 1) (c.f. Eq. (121)) and there is an overall − sign for O(N) (it would be + for Sp(2N) and 0 for SU(N)). Thus

$$A_4^{(\text{Moebius})} = -K_4 \int_0^1 \frac{dq}{q} \int_0^2 \prod_{I=1}^{3} d\nu_I \, \theta(\nu_{I+1} - \nu_I) \prod_{I<J} (\psi_{IJ,N})^{k_I \cdot k_J} \tag{174}$$

In fact, the annulus and the Moebius strip are so similar that we may write[36]

$$A_4^{(\text{Annulus})} + A_4^{(\text{Moebius})} = \int_0^1 \frac{dq}{q} \left[NF(q^2) - 8F(-\sqrt{q}) \right] \tag{175}$$

and changing variables to $\lambda = q^2$ and $\lambda = \sqrt{q}$ respectively in the two terms, this can be written for O(32) only as

$$16 \int_{-1}^{+1} \frac{d\lambda}{\lambda} F(\lambda) \qquad \text{(principal part)} \tag{176}$$

We should say that no consistent regularization has been worked through in detail for superstrings. Dimensional regularization is presumably not useful since d = 10 seems fixed. Thus, the above procedure is ambiguous: for example, if we put $\lambda = q$ instead of q^2 for the annulus the result would diverge. Nevertheless, the principal part prescription is the way to isolate O(32) and will be used in what follows.

Consider now the M = 5 pentagon (Fig. 26). The divergence cancellation for $M \geq 6$ will become clear from M = 5. Really the M = 4 case is too close to the nonrenormalization theorems for the general case to be clear.

In the vector emission vertex, Eq. (107), both P^i and R^{ij} have both zero modes (n = 0) and nonzero modes (n ≠ 0). P is linear in α_n while R is bilinear in S_n. We need at least eight S_o factors for

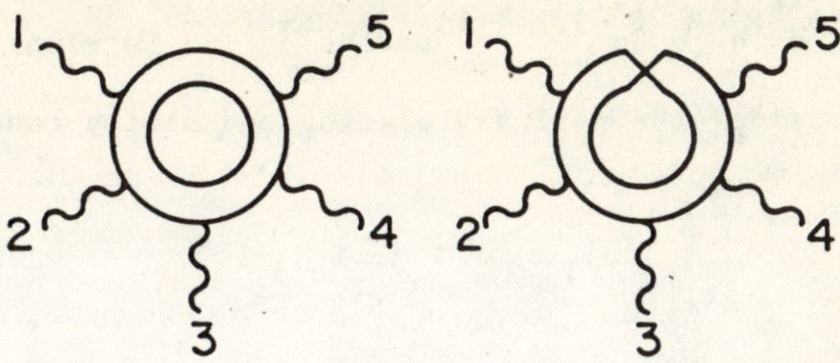

Figure 26

the trace to be nonvanishing.

Already for M = 4 we needed

$$t^{i_1j_1i_1j_2i_3j_3i_4j_4} = \mathrm{tr}\,(R_o^{i_1j_1} R_o^{i_2j_2} R_o^{i_3j_3} R_o^{i_4j_4}) \tag{177}$$

$$\begin{aligned} &= a_1 \varepsilon^{i_1j_1i_2j_2i_3j_3i_4j_4} \\ &+ a_2[(\delta^{i_1i_2}\delta^{j_1j_2} - \delta^{i_1j_2}\delta^{i_2j_1}) \\ &(\delta^{i_3i_4}\delta^{j_3j_4} - \delta^{i_3j_4}\delta^{i_4j_3}) \\ &+ (13)(24) + (14)(23)] \\ &+ a_3[\delta^{j_1i_2}\delta^{j_2i_3}\delta^{j_3i_4}\delta^{j_4i_1} + \cdots]_{48\ \text{terms}} \end{aligned} \tag{178}$$

The coefficients a_i of these three irreducible tensors can be found by special choices of indices e.g.

$$a_1 = \mathrm{tr}(R_o^{12} R_o^{24} R_o^{56} R_o^{78}) = -\frac{1}{32}\,\mathrm{tr}(\gamma^- \gamma^{78} \gamma^{56} \gamma^{34} \gamma^{12} h_- \gamma^+) \tag{179}$$

$$= -\frac{1}{2} \tag{180}$$

$$a_2 = \mathrm{tr}(R_o^{12} R_o^{12} R_o^{34} R_o^{34}) = -\frac{1}{32}\,\mathrm{tr}(\gamma^- \gamma^{34} \gamma^{34} \gamma^{12} \gamma^{12} h_- \gamma^+) \tag{181}$$

$$= -\frac{1}{2} \tag{182}$$

$$a_3 = \mathrm{tr}(R_o^{12} R_o^{23} R_o^{34} R_o^{41}) = (+1)_{\text{bosons}} + (-\tfrac{1}{2})_{\text{fermions}} = +\tfrac{1}{2} \quad (183)$$

For M = 5 we consider the parity-violating and parity-conserving pieces of

$$\mathrm{tr}\left(R_o^{i_1 j_1} R_o^{i_2 j_2} R_o^{i_3 j_3} R_o^{i_4 j_4} R_o^{i_5 j_5}\right) \quad (184)$$

The parity-violating part can be written

$$c_1[\delta^{i_1 j_2} \varepsilon^{j_1 i_2 j_3 i_4 j_4 i_5 j_5} + \cdots]_{20 \text{ terms}}$$

$$+ c_2[\delta^{i_1 j_3} \varepsilon^{j_1 i_3 i_2 j_2 i_4 j_4 i_5 j_5} +]_{20 \text{ terms}} \quad (185)$$

where the terms group into those with nearest neighbor and next-nearest neighbor vertices on ϕ respectively. By considering particular indices one finds $c_1 = 3/20$ and $c_2 = 1/20$. Actually, there are Gram determinental constraints which reduces the 40 tensors in (185) to 31. These constraints are the analog of

$$\delta^{\mu\alpha} \varepsilon^{\beta\gamma\delta\varepsilon} + \delta^{\mu\beta} \varepsilon^{\gamma\delta\varepsilon\alpha} + \delta^{\mu\gamma} \varepsilon^{\delta\varepsilon\alpha\beta} + \delta^{\mu\delta} \varepsilon^{\varepsilon\alpha\beta\gamma} + \delta^{\mu\varepsilon} \varepsilon^{\alpha\beta\gamma\delta} = 0 \quad (186)$$

in d = 4. But we may keep an overcomplete basis if we assign special indices and do not use tensor identities (i.e. do not compare the non-unique coefficients of specific tensors).

The parity-conserving pieces has 524 terms of the types

$$[(\delta\delta)(\delta\delta\delta)]_{160 \text{ terms}} + [(\delta\delta\delta\delta\delta)]_{384 \text{ terms}} \quad (187)$$

We may separate these further into irreducible parts by giving typical groupings:

$$a_1[(12)(345)]_{80 \text{ terms}} + a_2[(13)(245)]_{80 \text{ terms}}$$

$$+ \ b_1[(12345)]_{32 \text{ terms}} \ + \ b_2[(13524)]_{32 \text{ terms}}$$

$$+ \ b_3[(13254)]_{160 \text{ terms}} + b_4[(12354)]_{160 \text{ terms}} \quad (188)$$

Looking at particular values of the indices gives $a_1 = a_2 =$

$-b_2 = b_3 = -b_4 = -\frac{1}{4}$ and $b_1 = \frac{3}{4}$. Note that the only bosonic contribution to this supertrace is in b_1.

Thus the contribution to A_P (P = planar) is [37]

$$A_P^{(R_o^5)} = \zeta_1^{i_1}\zeta_2^{i_2}\zeta_3^{i_3}\zeta_4^{i_4}\zeta_5^{i_5}\, k_1^{j_1}k_2^{j_2}k_3^{j_3}k_4^{j_4}k_5^{j_5}$$

$$\mathrm{tr}(R_o^{i_1j_1}R_o^{i_2j_2}R_o^{i_3j_3}R_o^{i_4j_4}R_o^{i_5j_5})$$

$$\int \prod_{I=1}^{5} dx_I \frac{1}{w}\left(-\frac{2\pi}{\ell n w}\right) \prod_{1\leq I<J\leq 5} (\psi_{IJ})^{k_I\cdot k_J} \qquad (189)$$

To go from A_P to A_N (N = non-orientable Moebius strip) we replace w by $-w$ and ψ_{IJ} by $\psi_{N,IJ}$. This is the only piece of A_5 with ten S_o factors, the rest have eight S_o factors. Consider first the terms with one S_n and one S_{-n} $(n \neq 0)$. There are eight nonvanishing terms in the factor e.g.

$$\frac{1}{64}\,\mathrm{Tr}\{[S_b^a + \sum_{k\geq 1}(\bar{S}_k^a + \bar{S}_{-k}^a)](\gamma^{i_1j_1-})^{ab}\, [S_o^b + \sum_{\ell\geq 1}(S_\ell^b + S_{-\ell}^b)]$$

$$[\bar{S}_o^c + \sum_{m\geq 1}\bar{S}_m^c x_2^{-m} + \bar{S}_{-m}^c x_2^m)](\gamma^{i_2j_2-})^{cd}\, [S_o^d + \sum_{p\geq 1}(S_p^d x_2^{-p} + S_{-p}^d x_2^p)]$$

$$W^N F\, R_o^{i_3j_3} R_o^{i_4j_4} R_o^{i_5j_5}\}\,. \qquad (190)$$

Note that $\mathrm{tr}(\bar{S}_k\gamma^{i_1j_1-}S_{-k}) = 0$ because there is no second rank antisymmetric tensor in O(8). The eight nonvanishing terms in (190) are all equal. We must then add the ten inequivalent pairs of vertices and obtain[37]

$$A_p^{(S_o^8 S_n S_{-n})} = \zeta_1^{i_1}\zeta_2^{i_2}\zeta_3^{i_3}\zeta_4^{i_4}\zeta_5^{i_5}\, k_1^{j_1} k_2^{j_2} k_3^{j_3} k_4^{j_4} k_5^{j_5}\, \frac{1}{64}$$

$$\int \prod_{I=1}^{5} dx_I w^{-1}\left(-\frac{2\pi}{\ell n w}\right)^5 \prod_{1\leq I<J\leq 5} (\psi_{IJ})^{k_I\cdot k_J}$$

$$\sum_{\ell\geq 1} \frac{1}{1-w^\ell} \{[(\frac{w}{x_2})^\ell - x_2^\ell]$$

$$t^{i_1j_1i_2j_2ij}\, tr(R_o^{ij} R_o^{i_3j_3} R_o^{i_4j_4} R_o^{i_5j_5})$$

$$+ [(\frac{w}{x_2x_3})^\ell - (x_2x_3)^\ell]t^{i_1j_1i_3j_3ij}\, tr(R_o^{ij} R_o^{i_2j_2} R_o^{i_4j_4} R_o^{i_5j_5})$$

$$+ \text{cyclic permutations}\}_{2x5\text{ terms}} \tag{191}$$

where

$$t^{i_1j_1i_2j_2ij} = tr(\gamma^{i_1j_1}\gamma^{i_2j_2}\gamma^{ij}) \tag{192}$$

$$= 32[\delta_{ij_1}\delta_{i_2j}\delta_{i_1j_2} - \delta_{ij_1}\delta_{i_2j}\delta_{j_ij_2}$$

$$-\delta_{ij_1}\delta_{i_1i_2}\delta_{jj_2} + \delta_{ii_1}\delta_{i_2j_1}\delta_{jj_2} - \delta_{ii_2}\delta_{i_1j_2}\delta_{jj_1}$$

$$+\delta_{ii_2}\delta_{i_1j}\delta_{j_1j_2} + \delta_{ij_2}\delta_{i_1i_2}\delta_{jj_1} - \delta_{ij_2}\delta_{i_2j_1}\delta_{i_1j}] \tag{193}$$

We obtain $A_N^{(S_o^8 S_n S_{-n})}$ from $A_P^{(S_o^8 S_n S_{-n})}$ by replacing ψ_{IJ} by $\psi_{N,IJ}$, w by $-w$ and x_1 by $-x_1$.

The remaining terms are of the form $(R_o^4 P)$ with

$$P^i = p^i + \sum_{n\geq 1}\sqrt{n}\,(a_n^i + a_n^{i+}) \tag{194}$$

The zero mode piece (p^i) modifies the gaussian in the momentum integral already computed in Eqs. (153) and (156) above. The nonzero modes change the bosonic trace in Eq. (144) above. Modifying the operator O_I of Eq. (135) to

$$\hat{O}_I^i = P^{i_I}(u_I)O_I = \sqrt{n}\,(au_I^{-1} + a^+u_I)O_I \tag{195}$$

giving a modified trace

$$\hat{T} = \frac{1}{\pi}\int d^2z\,\sqrt{n}\,(z_1u_1^{-1} + z^*u_1)e^{-|z|^2}\,e^{-\beta_1}\,e^{z^*z_1} \tag{196}$$

$$= \sqrt{n}\left(Wu_i^{-1}\frac{\partial}{\partial B} + Cu_1^{-1} + u_1\frac{\partial}{\partial C}\right)T \tag{197}$$

where T is the old trace (c.f Eq. (144))

$$T = \frac{1}{1 - W}\exp\left(-\frac{D(W - 1) - BC}{1 - W}\right) \tag{198}$$

It follows that

$$\hat{T} = \frac{1}{1 - W}\left(Cu_1^{-1} + Bu_1\right)T \tag{199}$$

Collecting results we find that[37]

$$A_p^{(R_o^4P)} = \int \prod_{I=1}^{5} dx_I\, w^{-1}\left(-\frac{2\pi}{\ell nw}\right)^5 \prod_{1\leq I<J\leq 5}(\psi_{IJ})^{k_I\cdot k_J}$$

$$[(\zeta_1^{i_1}\sum_{L=1}^{5} k_L^{i_1}\frac{\ell n(x_1\cdots x_L)}{-\ell nw}K_4(2345)$$

$$+\zeta_2^{i_2}\sum_{L=2}^{1} k_L^{i_2}\frac{\ell n(x_2\cdots x_L)}{-\ell nw}K_4(3451)$$

$$+\text{ cyclic permutations})_{5\text{ terms}}$$

$$+\{K_4(2345)\zeta_1^{i_1}\sum_{\ell\geq 1}\frac{1}{1-w^\ell}(k_2^{i_1}[x_2^\ell - (\frac{w}{x_2})^\ell]$$

$$+k_3^{i_1}[(x_2x_3)^\ell - (\frac{w}{x_2x_3})^\ell]$$

$$+k_4^{i_1}[(\frac{w}{x_5x_1})^\ell - (x_5x_1)^\ell] + k_5^{i_1}[(\frac{w}{x_1}) - x_1^\ell])$$

$$+ \text{ cyclic permutations}\}_{5 \text{ terms}}] \tag{200}$$

where e.g.

$$K_4(2345) = \zeta_2^{i_2}\zeta_3^{i_3}\zeta_4^{i_4}\zeta_5^{i_5}\ k_2^{j_2}k_3^{j_3}k_4^{j_4}k_5^{j_5}\ \mathrm{tr}(R_o^{i_2 j_2} R_o^{i_3 j_3} R_o^{i_4 j_4} R_o^{i_5 j_5}) \tag{201}$$

As usual, for the nonorientable Moebius strip we replace ψ_{IJ}, w, x_1 by $\psi_{N,IJ}$, $-w$, $-x_1$ respectively.

This completes the M = 5 results. Higher point functions are computed similarly. We now make an asymptotic expansion for $q \to 0$

$$A_p = \int_0^1 \frac{dq}{q}\left[a_o + \frac{a_1}{\ell nq} + \frac{a_2}{(\ell nq)^2} + \cdots\right] \tag{202}$$

The infra-red divergence $(q \to 0)$ is from the a_o and a_1 terms only. From the physical picture of the dilaton pole, presumably all $a_k = 0$ for $k > 1$. We should note the (ℓnq) factors are $(\ell nq)^5$ and $(\ell nq)^{-M-1}$ from the momentum integrand and Jacobian respectively. for M = 4 it is trivially meromorphic and the annulus and Moebius strip diagrams (Fig. 27) cancelled. Recall that we put $q = q'^4$ to relate the q variables and used the change

$$\int_0^2 \prod_{I=1}^{M-1} d\nu_I' = 2^{M-1} \int_0^1 \prod_{I=1}^{M-1} d\nu_I \tag{203}$$

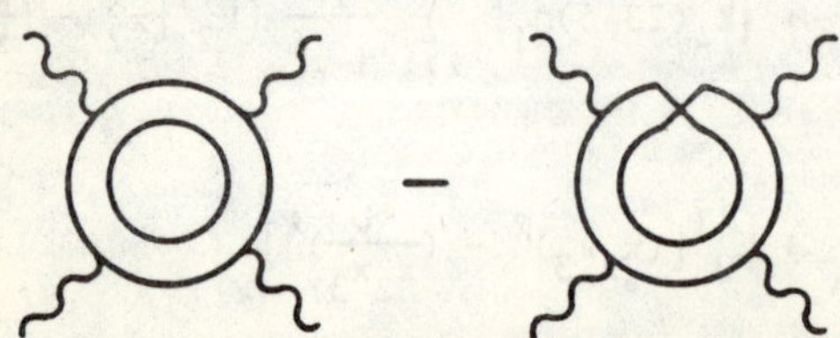

Figure 27

together with

$$\frac{dq}{q} = 4\,\frac{dq'}{q'} \tag{204}$$

to obtain (for M = 4) 4 times 8 = 32. But how can the 2^{M-1} be independent of M, the number of external lines? The answer lies in the mode sum

$$F_p(\nu, w) = \sum_{n=1}^{\infty} \frac{c^n - (w/c)^n}{1 - w^n} \quad \text{and } c = w^{\nu} \tag{205}$$

$$= \sum_{n=1}^{\infty} \sum_{m=o}^{\infty} (e^{n\nu\ell nw} - e^{n(1-\nu)\ell nw})e^{mn\ell n\nu} \tag{206}$$

$$= \sum_{m=0}^{\infty} \left(\frac{e^{(\nu+m)\ell nw}}{1 - e^{(\nu+m)\ell nw}} - \frac{e^{(1-\nu+m)\ell nw}}{1 - e^{(1-\nu+m)\ell nw}}\right) \tag{207}$$

$$= \frac{1}{(-\ell nw)} (A_o + A_1(-\ell nw) + A_2(-\ell nw)^2 + \cdots) \tag{208}$$

Now we find easily that

$$A_o = \frac{1}{\nu} + 2\nu \sum_{m=1}^{\infty} \frac{1}{\nu^2 - m^2} = \pi \cot \pi\nu \tag{209}$$

The nonorientable Moebius strip contains

$$F_N(\nu, w) = \sum_{\ell=1}^{\infty} \frac{1}{1 - (-w)^{\ell}} [c^{\ell} - (-w/c)^{\ell}] \tag{210}$$

$$\sim - \left(\frac{1}{2}\right) \frac{\ell nq}{2\pi} \cot\left(\frac{\pi\nu}{2}\right)\left[1 + 0\left(\frac{1}{\ell nq}\right)\right] \tag{211}$$

as $q \to 0$. Note that

$$F_p(\nu, w) \sim - \frac{\ell nq}{2\pi} \cot \pi\nu \left[1 + 0\left(\frac{1}{\ell nq}\right)\right] \tag{212}$$

the extra (1/2) in Eq. (211) is the essential point. When we generate the extra $(\ell nq)^{M-4}$ from the mode sums, necessary to identify the coefficient a_o in Eq. (202), we also generate a relative factor 2^{4-M}

in the Moebius strip relative to the annulus. Thus the terms may be combined as

$$\frac{1}{2} \int_0^1 \frac{d\lambda}{\lambda} \left[NF(\lambda) - \frac{4 \cdot 2^{M-1}}{2^{M-4}} F(-\lambda) \right] = 16 \int_{-1}^{+1} \frac{d\lambda}{\lambda} F(\lambda) \qquad (213)$$

and the principal part prescription generalizes to all M for the gauge group O(32).

References

1. P. H. Frampton, Dual Resonance Models, Benjamin (1974).

2. G. Veneziano, Nuovo Cim. 57A, 190 (1968).

3. K. Igi and S. Matsuda, Phys. Rev. Lett. 18, 625 (1967); Phys. Rev. 163, 1622 (1967).

4. S. Mandelstam, Phys. Rev. 166, 1539 (1968).

5. H. Harari, Phys. Rev. Lett. 22, 562 (1962).

6. J. L. Rosner, Phys. Rev. Lett. 22, 689 (1969).

7. P. G. O. Freund, Phys. Rev. Lett. 20, 235 (1968);
H. Harari, Phys. Rev. Lett. 20. 1395 (1968).

8. H. M. Chan and T. S. Tsun, Phys. Lett. 28B, 485 (1969);
K. Bardakci and H. Ruegg, Phys. Rev. 181, 1884 (1969);
Z. Koba and H. B. Nielsen, Nucl. Phys. B10, 633 (1969);
ibid. B12, 517 (1969).

9. S. Fubini and G. Veneziano, Nuovo Cim. 64A, 811 (1969);
K. Bardakci and S. Mandelstam, Phys. Ref. 184, 1640 (1969).

10. Y. Nambu, in Symmetries and Quark Models, editor R. Chand, p. 269 (1969).
S. Fubini, D. Gordon and G. Veneziano, Phys. Lett. 29B, 679 (1969).

11. Y. Nambu, lecture notes prepared for the Summer Institute of the Niels Bohr Institute, Copenhagen (1970);
T. Goto, Prog. Theor. Phys. 46, 1560 (1971);
P. Goddard, J. Goldstone, C. Rebbi and C. B. Thorn, Nucl. Phys. B56, 109 (1973).

12. D. J. Gross, A. Neveu, J. Scherk and J. H. Schwarz, Phys. Rev. D2, 697 (1970).

13. R. C. Brower, Phys. Rev. D6, 1655 (1972).
E. Del Guidice, P. Di Vecchia and S. Fubini, Ann. Phys. 70, 378 (1972);
P. H. Frampton and H. B. Nielsen, Nucl. Phys. B45, 318 (1972);
P. Goddard and C. B. Thorn, Phys. Lett. 40B, 235 (1972).

14. M. A. Virasoro, Phys. Rev. 177, 2309 (1969).

15. P. Ramond, Phys. Rev. D3, 2415 (1971);
A. Neveu and J. H. Schwarz, Nucl. Phys. B31, 86 (1971).

16. T. Yoneya, Prog. Theor. Phys. 51, 1907 (1974);
J. Scherk and J. H. Schwarz, Nucl. Phys. B81, 118 (1974); Phys. Lett. 57B, 463 (1975).

17. D. J. Gross, J. A. Harvey, E. Martinec and R. Rohm, Phys. Rev. Lett. 54, 502 (1985); Nucl. Phys. B256, 253 (1985) and to be published.

18. M. B. Green and J. H. Schwarz, Phys. Lett. 109B, 444 (1982).

19. P. H. Frampton, Phys. Lett. 122B, 351 (1983);
P. H. Frampton and T. W. Kephart, Phys. Rev. Lett. 50, 1343, 1347 (1983); Phys. Rev. D28, 1010 (1983).

20. P. H. Frampton and T. W. Kephart, Phys. Lett. 131B, 80 (1983).

21. P. H. Frampton, Phys. Lett. 140B, 313 (1984).

22. L. Caneschi, G. Farrar and A. Schwimmer, Phys. Lett. 138B, 386 (1984).

23. P. H. Frampton and K. Yamamoto, Phys. Rev. Lett. 52, 2016 (1984).

24. P. H. Frampton and K. Yamamoto, Nucl. Phys. 254B, 349 (1985).

25. S. Randjbar-Daemi, A. Salam and J. Strathdee, Phys. Lett 124B, 345 (1983).
P. H. Frampton, P. J. Moxhay and K. Yamamoto, Phys. Lett. 144B, 354 (1984).

26. S. Randjbar-Daemi, A. Salam and J. Strathdee, Nucl. Phys. B214, 491 (1983).

27. S. Randjbar-Daemi, A. Salam and J. Strathdee, Phys. Lett. 132B, 386 (1984); Nucl. Phys. B242, 447 (1984).

28. H. Georgi, Nucl. Phys. B156, 126 (1979).

29. P. H. Frampton, Phys. Lett. 89B, 352 (1980).

30. M. B. Green and J. H. Schwarz, Phys. Lett. 149B, 117 (1984).

31. G. F. Chapline and N. S. Manton, Phys. Lett. 120B, 105 (1983). E. Bergshoeff, M. De Roo, B. De Wit and P. Van Nieuwenhuizen, Nucl. Phys. B195, 97 (1982).

32. L. Alvarez-Gaume and E. Witten, Nucl. Phys. B234, 269 (1983).

33. P. H. Frampton and K. Yamamoto, Phys. Lett. 156B, 345 (1985).

34. J. H. Schwarz, Phys. Reports 89, 223 (1982). This review is more current than Reference 1 and is a good source for more background material.

35. D. Amati, C. Bouchiat and J. L. Gervais, Nuovo Cim. Lett. 2, 399 (1969).

36. M. B. Green and J. H. Schwarz, Phys. Lett 151B, 21 (1985).

37. P. H. Frampton, P. Moxhay and Y. J. Ng, UNC-Chapel Hill Reports IFP-255-UNC, to be published in Phys. Rev. Lett. and IFP-256-UNC.

INTRODUCTION TO THE FIELD THEORY OF STRINGS

MICHIO KAKU

City College of the City University
of New York, N.Y. 10031

Lectures Presented at the Superstring Workshop,
Lewes Center for Physics, University of Delaware,
July 8-12, 1985

ABSTRACT

Years ago, we presented the field theory of interacting strings in the light cone gauge. Today, however, there is renewed interest in writing down the full gauge covariant field theory of strings. Such a construction may prove crucial to a non-perturbative solution of the outstanding problem of string theory: breaking the theory from 26 or 10 dimensions down to four dimensions.

1. Introduction

Although string theory is a mathematical framework of unparalleled elegance and beauty, a beginner learning string theory for the first time often goes through a frustrating and painful process. To a student who has been trained to systematically derive all the physical properties of a theory from a single field theory action, string theory, at best, seems to be a loose collection of folklore and rules-of-thumb. At worst, the theory may seem artificial or even contrived.

Furthermore, although string theory is often adverised as a candidate for a "theory of everything," one is disappointed that the theory does not manifestly contain general covariance and O(N) gauge symmetry. It is by "magic" that general covariance pops out of the theory in the zero slope limit. It seems to be purely an accident that the theory contains a spin two graviton possessing general covariance.

There are two reasons for this unfortunate state of affairs.

a) historically, the model has been evolving backwards for the past 15 years. The accidental discovery of the beta function reversed the normal historical evolution that characterizes most theories. The beta function became an "answer looking for a question." Because of this, string theory has perhaps the most unusual history of any physical theory, with the most bizarre twists and turns. And because the model "died" around 1974 as a theory of hadrons, its development was prematurely terminated.

b) because the model was discovered backwards, it was formulated as a first quantized theory, rather than as a genuine second quantized field theory. First quantization is perhaps the clumsiest framework in which to formulate a theory. It is notoriously difficult, for example, to discuss gauge invariance within a first quantized formalism.

Normally, one starts with a field theory action which manifestly contains all the gauge symmetries of the theory. From this action, one can derive the sum total of all information from the theory. However, the string model is not usually formulated as a field theory. The Nambu action is a first quantized action which elegantly describes the spectrum of a free string, but suffers from all the standard defects of first quantized theories when generalized to interactions. The problems of the first quantized formalism are:

a) interactions must be put in "by hand" in first quantized theories by summing over different topologies. (The first quantized action does not contain any interaction terms at all.)

b) the counting of perturbation diagrams must also be done by hand. Unitarity is restored only by carefully adjusting the weights of various perturbation diagrams to satisfy the unitarity condition. Unitarity does not emerge from a hermitian hamiltonian.

c) the origin of local gauge principles, such as general covariance, is obscure in a first quantized approach.

d) the theory is necessarily formulated perturbatively, i.e. it is expanded around free strings. It is not clear how to apply non-perturbative methods to the theory.

e) the Nambu action suffers from the problem that the perturbation series is dual, meaning that the sum over s channel Regge poles equals the sum over t channel poles. In usual field theories, we add the contributions over the two channels separately.

However, if we had a second quantized field theory, then the action would explicitly contain all interaction terms, the weights of all diagrams would be fixed, and unitarity would be automatic. Furthermore, the action should be invariant under some gauge principle which would contain general covariance and O(N) invariance as a subset.

Unfortunately, this is not presently the case. As stressed by Witten [1], Einstein discovered the general theory of relativity because he first started with a fundamental principle (general covariance) and then demanded that his action be invariant under this principle. Similarly, Yang and Mills began with local SU(2) symmetry and demanded that the action be invariant under this symmetry. However, the string model suffers from the embarrassing problem that the complete field theory action is known (in the light cone gauge) but its underlying gauge principle is at the present time completely unknown. This is the reverse of the usual historical process.

Years ago, we [2] wrote down the field theory of interacting strings in the light cone gauge. Because it was a second quantized theory, all the known properties of the string could be derived from one action principle, rather than appealing to folklore. But because we did not understand how to incorporate genuine gauge invariance into the theory, we had to eliminate all the redundant ghost modes by quantizing the theory in the light cone gauge. Although we were able to cataloguc all possible field theory interaction terms in the action, we made no pretense that we understood the gauge principle behind string theory.

At that time, of course, we realized that the true gauge symmetry behind string field theory must be a fantastically elegant one, because it contained general covariance and O(N) gauge symmetry as a subset, but we did not have any clues about how to begin such a formulation.

In fact, one could argue that our second quantized field theory only reproduced the old Veneziano model and nothing more. Of course, one could show that the perturbation theory was unitary because it was now derived from a hermitian hamiltonian, but one could also argue, with considerable justification, that it gave us no new information about the theory.

1. E. Witten, quoted in *Physics Today*, July, 1985.
2. M. Kaku and K. Kikkawa, Phys. Rev. D10 (1974) 1110, 1823.

Unfortunately, shortly after the light cone field theory was written down, string theory "died". The fact that the theory was defined only in 26 or 10 dimensions was the kiss of death. String field theory was only half complete when the theory died. Only the bravest continued to work on a theory which seemed to have only a glimmer of hope.

However, the recent "rebirth" of string theory due to the work of Green and Schwarz [3] has revived interest in writing a gauge covariant field theory of strings.

There is, moreover, now a pressing physical motivation behind writing down such a gauge covariant theory: to solve the outstanding problem in string theory, which is how to break 26 or 10 dimensions down to 4 dimensions. Only then can we make contact with physically measureable parameters and experimentally test the model. However, because the model is so tightly interwoven, one cannot simply "add Higgs" to the theory at will. It seems that only a dynamical, non-perturbative approach to the model will yield symmetry breaking.

However, our arsenal of non-perturbative techniques developed in the last decade (e.g. instantons, lattices, etc.) all depend crucially on the geometric and topological properties of gauge theories. It seems likely, therefore, that dynamical symmetry breaking in string theory will necessarily depend on a detailed understanding of the gauge principle behind the theory. We see, therefore, that it is not simply a mathematical exercise to discover the gauge principle behind interacting strings.

(The alternative point of view is to use perturbation theory to write down an "effective action" based on summing over loops. This effective action approach [4] in principle should be able to decide which vacuums are stable and which are not. Unfortunately, this method suffers from several problems. First, the expansion parameter is the slope. Because the size of the alternate six dimensional universe is also of that order, there will be intricate mixing of orders which will make the theory much more difficult to interpret. Second, it is still a perturbation theory, so we can only determine local minima. Non-perturbative effects proportional to $\exp - g^{-2}$ will never be seen.)

3. M. Green and J. H. Schwarz, Phys. Lett. 149B (1984) 117.
4. E.S. Fradkin and A.A. Tseytlin, Lebedev Inst., N 26I (1984); C.G. Callan, E.J. Martinec, M.J. Perry and D. Friedan, Princeton Univ. preprint (1985); K. Kikkawa and M. Yamasaki, Osaka preprint, OU-HET 83.

Because of renewed interest in string theory, the problem of extending the old field theory of strings to the gauge covariant case has resulted in a flurry of papers. Recently, we [5] improved on our original action by writing down a free string action which is manifestly invariant under gauge transformations. We also generalized this action to include the superstring [6] and the Princeton heterotic string [7]. We hope shortly to write down the complete interacting theory, from which we should be able to abstract the gauge principle.

It is our feeling that such a gauge covariant construction will prove crucial in settling the question of how to break 26 or 10 dimensions down to 4.

We should stress, however, that even when we construct the entire interacting gauge theory, this does not mean that we understand the underlying geometry behind the theory. For example, a field theorist, when writing down a theory of interacting gravitons, would simply power expand the theory of a massless spin 2 two particle (which is unique) in terms of Newton's constant. However, because the action is power expanded around free gravitons, the field theorist would never see the beautiful geometry behind the model and would miss completely the black hole, the big bang, and other classical solutions. On the other hand, a geometer would simply write down curvature scalar R as the action of the theory.

In other words, once we have completed the gauge theory of interacting strings, our work is not yet over. We must realize that the theory would be a power expansion around free strings, much like a theory of interacting spin 2 massless fields is a power expansion of Einstein's theory around free particles. We must still sum the series to find the counterpart of Einstein's action $\sqrt{-g}\,R_{\mu\nu}g^{\mu\nu}$ for the theory of strings.

5. M. Kaku, "Gauge Field Theory of Covariant String," to appear in Nucl. Phys. This action was simultaneously proposed by T. Banks and M. Peskin, see Proceedings of the Argonne Symposium on Anomalies, Geometry, and Topology, March, 1985, World Scientific.

6. M. Kaku and J. Lykken, "Supergauge Field Theory of Superstrings", Proceedings of the Argonne Symposium on Anomalies, Geometry, and Topology, 1985.

7. M. Kaku, "Supergauge Field Theory of Heterotic Covariant Strings, Osaka preprint, OU-HET 79, Aug. 1985.

2. From First to Second Quantization

Let us begin by reviewing some basic facts about making the transition from first to second quantized theories.

The hallmark of a first quantized theory is that it is formulated in terms of the co-ordinate X_μ which labels the location of a point particle or a string. The canonical commutation relations are then defined in terms of the co-ordinates and their conjugates. By contrast, a second quantized field theory is formulated in terms of a field functional $\Phi(X)$ for a point particle and $\Phi(X_1, X_2, X_3, \cdots)$ for a string, where the subscript refers to the various normal modes of the string.

Feynman [8] years ago showed us that the Green's functions of a theory can be formulated either in first or in second quantized language. If a non-relativistic point particle travels from point X and moves to Y, then Feynman showed that we can represent the Green's function in terms of a path integral where the weight of each path is given by the exponential of an action given by $\frac{1}{2}mv^2$. He then showed that the Green's functions could also be written down by minimizing a path integral defined in terms a second quantized theory with an action given by the Schrodinger equation.

We begin with the Green's function connecting a particle that originally started at point X_i and time t_i and then propagates to X_j and time t_i . We will eventually slice up the interval into finite intervals i, i+1, i+2,...j. This Green's function $G(X_i, X_j)$ can be written either in first or in second quantized formalisms, i.e. either with the position vector X_i as the independent variable, or with a field functional $\Phi(X)$ composed of all values of the position vector. We will use the following formula to *derive* an expression for the second quantized action $L(\Phi)$ when we know the first quantized action $L(X)$:

$$\int_{X_i}^{X_j} DX \exp i \int L(X)\, d\sigma d\tau = \int D\Phi^\dagger(X)\, D\Phi(X)\;\; \Phi^\dagger(X_i)\, \Phi(X_j) \exp i \int L(\Phi) DX$$

If we take the Lagrangian to be $\frac{1}{2}m\dot{X}_i^2$, then this expression can be explicitly evaluated by actually performing all Gaussian integrals. We find that this Green's function is equal to:

$$[2\pi i(t_j - t_i)/\, m]^{-\frac{1}{2}} \exp \tfrac{1}{2} im(X_j - X_i)^2/\,(t_j - t_i)$$

---

8. R.P. Feynman and A.R. Hibbs, *Quantum Mechanics and Path Integrals*, McGraw-Hill Book Co., N.Y. 1965.

This expression can be viewed as a matrix connecting variables defined at point j with variables at point i. As a result, we can use the Gaussian integral:

$$(\det A)^{-1} = \int D\Phi_n \, D\Phi_m^* \exp -[\Phi_n A_{nm} \Phi_m^*]$$

where n and m label an infinite number of parameters. Using the observation that:

$$G(X_j, X_i) = \left[i\,\partial_t - \tfrac{1}{2}m \nabla^2 \right]_{ij}^{-1}$$

we now can show that:

$$L(\Phi) = \Phi(X)^* \left[i\,\partial_t - \tfrac{1}{2}m \nabla^2 \right] \Phi(X)$$

Notice that we have now smoothly made the transition from the first quantized to the second quantized action, deriving the Schrodinger equation from the action of a point particle $\frac{1}{2}m\dot{X}_i^2$.

Note that the first quantized action $\frac{1}{2}mv^2$, by itself, contains no interaction terms. However, we can introduce interactions if we integrate the path integral over a Y-shaped region in the first quantized formalism. This easily translates, in the second quantized language, to adding an explicit Ψ^3 term in our field theory action.

The lesson here is that, although the first quantized action $\frac{1}{2}mv^2$ contains no interaction term, interactions are added by hand by minimizing the action over a Y-shaped topology.

These results can be easily carried over to the relativistic case. If we start with the action $\sqrt{\dot{X}_\mu^2}$ then we can write down the Lagrangian as follows:

$$L = \int_{t_1}^{t_2} m \sqrt{\dot{X}_\mu^2}$$

Notice that this action is invariant under a simple reparametrization of the "time" variable τ. Following Dirac, one can write down the first class constraints of this action:

$$\frac{\delta L}{\delta \dot{X}_\mu} \equiv P_\mu; \quad P_\mu^2 - m^2 = 0$$

This first class constraint can be incorporated into our path integral by going to the second order formalism by adding two new fields, the field conjugate to X and a Lagrange multiplier:

$$L = P\dot{X} + \lambda[P^2 - m^2]$$

Notice that our naive hamiltonian is formally zero, which is typical of gauge theories. In order to introduce an effective hamiltonian let us choose the gauge λ = 1. Our first quantized action now reads :

$$L = P\dot{X} + P^2 - m^2$$

Inserting this into our path integral, we can now extract out the second quantized action $L(\Phi)$. This is most easily done by taking infinitesimal "time" slices in the path integral, and then converting the term $P\dot{X}$ into a series of intermediate states, defined from i to j. Labeling each τ slice by i, i+1, i+2,,,. we can now make the transition to quantum operators for each slice:

$$\int_{X_i}^{X_j} DP \exp i \int d\sigma d\tau \, P_\mu \dot{X}^\mu \rightarrow \int \prod_{n=i}^{j-1} DX_n DP_n \exp i \int d\sigma P_{i,\mu} (X_i^\mu - X_{i+1}^\mu)$$

$$= \int \prod_{k=i}^{j-1} \langle X_k | P_k \rangle DX_k \; DP_k \; \langle P_k | X_{k+1} \rangle$$

Notice that our new action for the relativistic scalar field now reads:

$$L = \Phi(X)^* [\, i \partial_\tau + \partial^2 + m^2 \,] \Phi(X)$$

Of course, the proper time τ is merely an artifact of the parametrization process and should not appear in the final theory. No physics can depend on a choice of parametrization in a parametrization independent theory. This τ can be eliminated in two ways. First, we can take the Fourier transform of the action. Because the field Φ has a trivial τ dependence, we can set $\Phi(X,\tau) = \exp iE\tau \;\; \Phi(X)$ so we can eliminate the E dependence by absorbing this constant into the mass term in the action. Second [9], we can analyze the Green's functions of the theory and then perform the integration over τ :

$$G(X_1,\tau_1;X_2,\tau_2) \rightarrow (\tau_2 - \tau_1)^{-2} \exp \left[\tfrac{1}{2} im(X_2 - X_1)^2 / (\tau_2 - \tau_1) \right]$$

Integrating out this expression over τ , we find:

$$\langle X_2 | [P^2 - m^2]^{-1} | X_1 \rangle$$

9. R. Casalbuoni, J. Gomis, and G. Longhi, Nuovo Cim. 24A (1974) 249.

In this way, we retrieve the usual Klein Gordon equation as a by-product of the path integral formalism.

The lessons derived from these trivial examples are simple

(a) a second quantized theory can be derived from the first quantized theory through path integrals, although complications arise due to gauge invariance

(b) interactions are introduced in a first quantized theory by summing over different topologies; interactions are introduced in a second quantized theory through terms in the action.

3. Light Cone Field Theory of Strings

This can now be generalized to the case of the Nambu string [10]. Let define the first quantized action to be the area swept out by a string moving in 26 dimensional space. Our path integral is now defined as: $L(X) = \sqrt{g_{01}^2 - g_{00}g_{11}}$ and $g_{00} = \dot{X}^2$; $g_{01} = \dot{X}X'$ and $g_{11} = X'^2$. The application of functional methods to the string model is straightfoward. Unlike the action for the non-relativistic point particle, this action is manifestly invariant under re-parametrizations of the two-dimensional surface swept out by the string $\sigma \to \sigma'(\sigma,\tau)$, $\tau \to \tau'(\sigma,\tau)$. This gauge symmetry necessarily implies that there are Dirac first class constraints in the theory. If we define the conjugate momentum

$$P_\mu \equiv \frac{\delta L}{\delta \dot{X}^\mu} = (-g)^{-\frac{1}{2}}[g_{01}X'_\mu - g_{11}\dot{X}_\mu]$$

then our first class constraints read:

$$P_\mu^2 + (\partial_\sigma X_\mu)^2 \equiv 0; \quad P_\mu \, \partial_\sigma X^\mu \equiv 0$$

Converting to the second order formalism, we can show that this Green's function can be written as:

$$G(X_i, X_j) = \int_{X_i}^{X_j} DX \, DP \, D\lambda \, D\rho \, \Delta \, \exp i \, \{ \int L(P,X) \, d\sigma d\tau \}$$

$$L(P,X) = P\dot{X} + (\tfrac{1}{2}\lambda)(P^2 + X'^2) + \rho(PX')$$

(Δ represents the measure and possible Fadeev-Popov determinants, which we keep unspecified.) If we functionall integrate over λ, ρ, and P, then we arrive back

10. Y. Nambu, Lectures at the Copenhagen Summer Symposium (1970).

at the original Nambu-Goto action. This second order action is explicitly invariant under a local gauge rotation with parameters $\epsilon(\sigma)$ and $\eta(\sigma)$:

$$\delta X_\mu = \epsilon P_\mu + \eta X_\mu' : \quad \delta P_\mu = \partial_\sigma (\epsilon X_\mu' + \eta P_\mu)$$

$$d\lambda = -\dot{\epsilon} + \lambda'\eta - \eta'\lambda + \rho'\epsilon - \epsilon'\rho$$

$$\delta\rho = -\dot{\eta} + \epsilon'\lambda - \lambda'\epsilon + \rho'\eta - \eta'\rho$$

This local gauge transformation guarantees that we can eliminate all ghosts.

Our starting point for the discussion of the second quantized field theory is now the generalized path integral [11] defined over surfaces with L loops and N external lines:

$$A_N = \sum_L \int_{T_L} \prod_i dz_i \; M_{N,L} \; \Delta_{FP} \exp \left| i \int d\sigma d\tau \sqrt{-g} + i \sum_j k_j X_j \right|$$

where M represents a non-trivial measure term corresponding to each topology, z_i represent the Koba-Nielsen variables defined either on the perimeter of a disk or on the real axis for open string amplitudes and on spheres for closed string amplitudes, and the sum over j represents the external momenta. Notice that we must sum over all possible topologies, i.e. closed and open surfaces with M handles, both orientable and non-orientable. [12]. (Fortunately, the solution of Poisson's equation on disks and spheres with M handles was worked out by Burnside in the 1890's, so we have, in principle, a complete understanding of the M-loop bosonic amplitude [13].)

From this expression for first quantized amplitudes, we will extract the entire field theory of strings defined in the light cone gauge.

In order to make the transition from the upper half plane to the string configuration, Mandelstam [14] has chosen the following conformal transformation:

$$\rho = \sum_i \alpha_i \log(z - z_i)$$

11. C.S. Hsue, B. Sakita, M.A. Virasoro, Phys. Rev. D2 (1970) 287; J.L. Gervais and B. Sakita, Phys. Rev. Lett. 30 (1973) 716.
12. K. Kikkawa, B. Sakita, M.A. Virasoro, Phys. Rev. 184 (1970) 1701.
13. M. Kaku and L.P. Yu, Phys. Rev. D3 (1971) 2992, 3007, 3020; M. Kaku and L.P. Yu, Phys. Lett. 33B (1970) 166; C. Lovelace, Phys. Lett. 32B (1970) 703; V. Alessandrini, Nuovo Cimento 2A (1971) 321; V. Alessandrini and D. Amati, Nuovo Cimento 4A (1971) 793.
14. S. Mandelstam, Nucl. Phys. B64 (1973) 205.

With this transformation, the Koba Nielsen variables, which were formally defined on a disk, are now mapped into the points at which the string splits and recombines.

Let us now chose the light cone gauge [15] where we choose X_+ to be proportional to τ . By solving for the first class constraints, we find that the Lagrangian of our theory is now just $P\dot{X}$, such that:

$$L = P\dot{X} = P_- \dot{X}_+ - P_i \dot{X}_i$$

where the i represents an index for the transverse states.

In other words, the hamiltonian of our theory is just P_- . All the longitudial states can be expressed entirely in terms of the transverse states.

The extraction of the propagator for the second quantized theory is perhaps the easiest exercise. Let us first consider the propagation of a string from one configuration to the next. Using precisely the same techniques developed earlier, we can now, by inspection, write down the second quantized action:

$$L_o = \Phi^\dagger(X,\tau) \; [i\,\partial_\tau - L_o + 1] \; \Phi(X,\tau)$$

where the theory is written entirely in terms of the physical transverse modes of the string.

We can now extract the interaction term for the theory in the same way that we extracted the Ψ^3 interaction term for the Schrodinger theory. Let us carefully slice the following region of the light cone diagram:

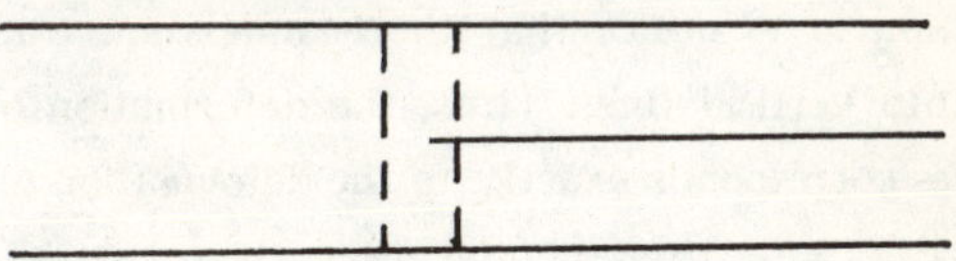

Because of the factorization of path integrals and the completeness theorem, we can extract from the path integral the following second quantized term for strings of length $\pi\alpha$:

$$L_3 = \Phi^\dagger(X_3)\Phi(X_1)\Phi(X_2) \; \delta[X_3(\sigma) - \Theta(\pi\alpha_1 - \sigma)X_1(\sigma) - \Theta(\sigma - \pi\alpha_1)X_2(\sigma)]$$

15. J. Goldstone, P. Goddard, C. Rebbi, and C. Thorn, Nucl. Phys. B56 (1973) 109.

where $0 < \sigma_3 < \pi\alpha_3$; $0 < \sigma_1 < \pi\alpha_1$; $\pi\alpha_1 < \sigma_2 < \pi\alpha_3$; $\sum_i \alpha_i = 0$.

It is gratifying to know that the string model prefers the simplest possible interaction term (the Dirac delta function) as the contact term for the splitting of strings.

At this point, we must now begin the process of extracting from the Mandelstam transformation all possible interactions of strings. This may seem like an arduous task, but the solution is actually quite simple if we exploit conformal invariance.

The key to understanding the splitting of strings is to notice that the first quantized action for the theory is conformally invariant. The problem of solving for the integrand of the amplitudes, therefore, reduces to the problem of finding the solution to Poisson's equation on disks.

In first year physics, we are taught to write down the equipotential lines on a given surface with charges. Lets start with a disk with four charges and draw these lines:

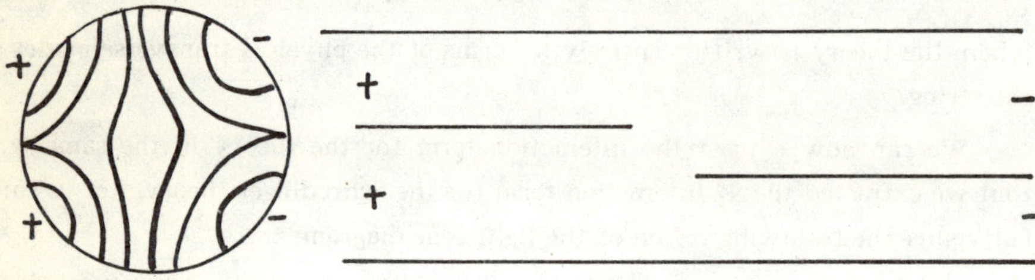

The crucial observation is to notice that the Mandelstam transformation maps equipotential lines into vertical lines. Thus, the deformation of the topology of the equipotential lines corresponds exactly to the deformation of moving strings.

In this way, we see immediately that a four-string term must be added to our action. If we place the charges in reverse order, we find:

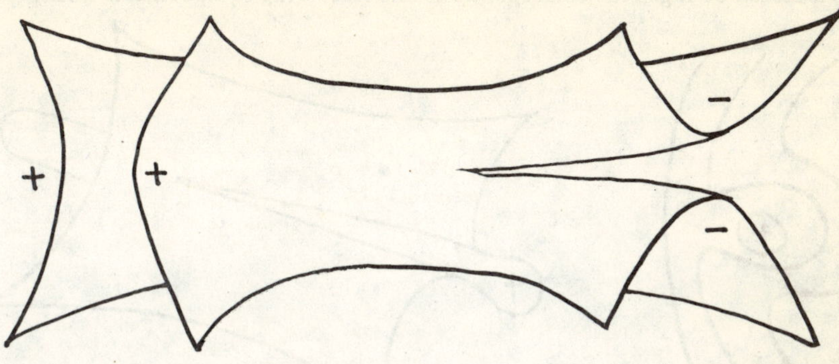

In this way, we can rigorously catalogue all possible interactions of strings. Closed strings, for example, correspond to charges placed in the center of disks or spheres. Obviously, the equipotential lines of the charges are just circles surrounding the charge. The three-tube interaction, which contains graviton-graviton interactions, can be expressed as the combining of two equipotential lines into one:

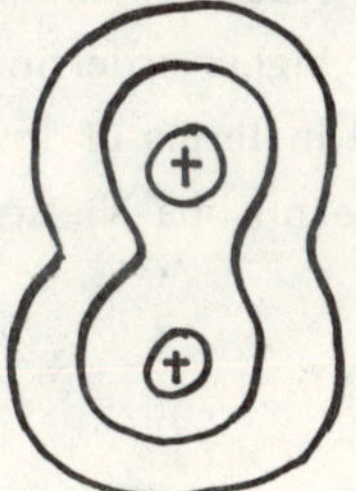

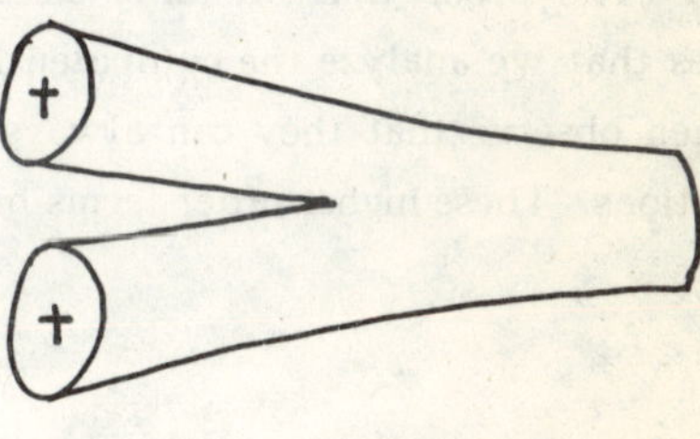

We also find that open strings can transform into closed strings. If we place a charge in the center of the diagram we then find that closed lines can touch the boundary of the disk and turn into open lines:

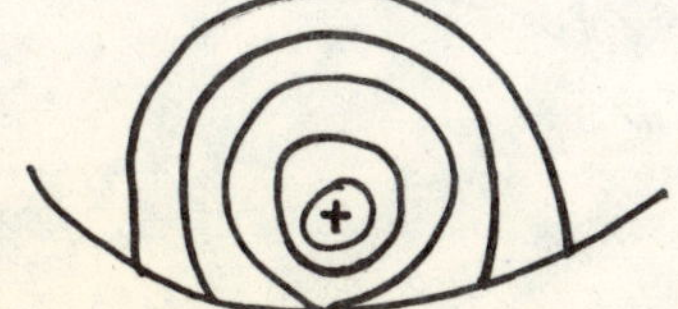

We also find that open strings can pinch with themselves at the center and peel off a closed string. This corresponds to the following equipotential line:

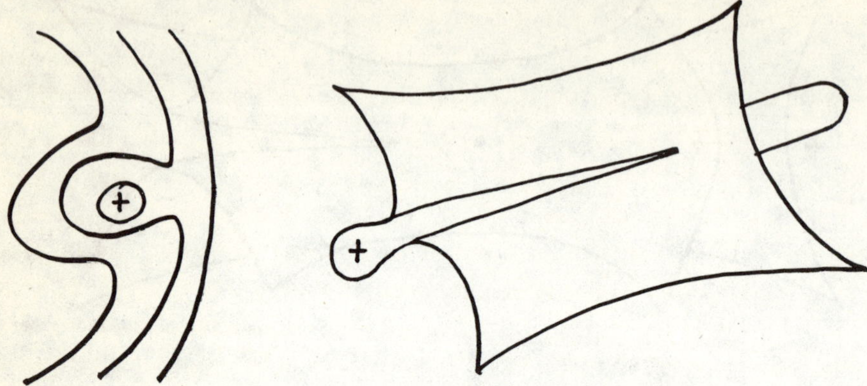

If we begin with a theory of open strings, we can show that higher string interaction terms can always be reduced back to these 5 basic interactions, including closed strings. Closed string theory, however, forms a complete set only with the three-tube interaction.

In this way, we can extract all five interaction terms for the light cone field theory. (The proof that we only need five string interactions and not more requires that we analyze the equipotential diagrams for higher order interactions and then observe that they can always be decomposed in terms of lower order interactions. These higher order terms have zero measure in Koba-Nielsen space.)

We stress that, although our arguments seem intuitive and pictoral, the results are mathematically rigorous because there is a one-to-one correspondence between equipotential lines and light cone diagrams, and there is a one-to-one correspondence between the first quantized path integral defined over these regions and terms generated by the second quantized action. See [2] for details.

Furthermore, because our action is *multilocal* rather than non-local, we see that our string field theory evades the problems of non-local field theory which have plagued field theorists for so many decades. Physicists who worked hard on non-local field theories, like Yukawa and Heisenberg, all grappled with the problem of causality. If we start with a non-local theory defined at two points X and Y, then an interaction at point X would inevitably create a disturbance at point Y faster than the speed of light. Even if the theory could be rigged so that the problem of causality could be superficially solved, the theory usually breaks down somewhere else.

One solution to this problem (perhaps the only solution!) is to place a series of points X_i close together along a line, so that the disturbance at one point will travel along the line and not exceed the speed of light. Furthermore, the points along this string can interact with another string only by instantaneously deforming its topology. In this way, string field theory does not violate Lorentz invariance.

4. String Field Theory in the Conformal Gauge

Fortunately, we have a complete understanding of the measure M in the light cone gauge necessary to make the transition from Koba-Nielsen variables to string interaction variables [16]. However, the measure M is not well understood for the conformal gauge, where we impose the choice $\lambda=1$ and $\rho=0$. As a result, our understanding of the Lorentz covariant conformal string theory is less well-developed at the level of interactions.

Let us begin a discussion of conformal string theory by starting with the path integral. Notice that, because we are fixing the gauge, there must be a Fadeev-Popov term. When we calculate the Fadeev-Popov ghost associated with this gauge, we find a new term in our action:

16. E. Cremmer and J.L. Gervais, Nucl. Physics B90 (1975) 410.

$$\Delta_{FP} \int d\epsilon d\eta \; \delta[\lambda(\epsilon,\eta)-1] \; \delta[\rho(\epsilon,\eta)] \equiv 1$$

$$\Delta_{FP} \equiv \det \begin{vmatrix} \frac{\delta\lambda}{\delta\epsilon} & \frac{\delta\rho}{\delta\epsilon} \\ \frac{\delta\lambda}{\delta\eta} & \frac{\delta\rho}{\delta\eta} \end{vmatrix}$$

$$= \det \begin{vmatrix} -\partial_\tau & \partial_\sigma \\ \partial_\sigma & \partial_\tau \end{vmatrix}$$

$$= \int d\theta \; dp_\theta \; \exp -i \left[\int d\sigma d\tau L(\theta, p_\theta) \right]$$

where $L(\theta, p_\theta)$ is given by:

$$p_\theta^1 (\partial_\sigma + \partial_\tau) \theta^1 + p_\theta^2 (\partial_\tau - \partial_\sigma) \theta^2$$

We, find, therefore, that the net effect of chosing the conformal gauge is to allow two ghosts to propagate in the theory. The two Fadeev Popov ghosts associated with the conformal gauge effectively *reduce the 26 dimensional excitations of the string down to 24 excitations.* Because the ghost particle now contains a τ derivative in the action, we must also let it contribute to the bra and ket vectors when we time slice the functional. Our Green's function now becomes $G(X_i, \theta_i; X_{i+1}, \theta_{i+1})$ which can be rewritten as:

$$< X_i \; \theta_i \,|\, P_i \; p_{\theta i} > \int DP_i \; Dp_{\theta i} < P_i \; p_{\theta i} \,|\, X_{i+1} \; \theta_{i+1} > \exp i \int d\sigma \left[P_i^2 + (\partial_\sigma X)^2 + p_{\theta i} \rho \partial_\sigma \theta_i \right]$$

where we have combined the spinors into one two dimensional spinor and ρ is the Pauli spin matrix σ_z . In order to actually perform the integration over the canonical momenta, we will find it useful to first calculate the matrix element between two ordinary string states:

$$< X_i \,|\, X_j > = \prod_n^\infty \delta(X_{ni} - X_{nj}) = < X_j \,|\, \Phi > \int D\Phi_n^* \; D\Phi_n \; \exp - \sum_n^\infty \Phi_n^* \; \Phi_n \; < \Phi \,|\, X_j >$$

$$= \int D\Phi^\dagger(X) \, D\Phi(X) \;\; \Phi(X_i)^\dagger \, \Phi(X_j) \exp - \int DX \;\; \Phi^\dagger(X) \, \Phi(X)$$

Now modify this matrix element by inserting $e^{iH\Delta\tau}$ between the state vectors:

$$< X_i \,|\, e^{iH\Delta\tau} \,|\, X_{i+1} > = \int D\Phi^\dagger(X) D\Phi(X) \; \Phi(X_i)^\dagger \Phi(X_{i+1})$$

$$\exp i \int DX \left\{ \Phi^\dagger(X_i) [\Phi(X_i) - \Phi(X_{i+1})] + i \, \Phi(X_{i+1})^\dagger H(X_{i+1}) \Phi(X_{i+1}) \Delta\tau \right\}$$

Notice that in the continuum limit, this expression reduces to the Lagrangian for the Schrodinger equation for an infinite number of free fields: $L = \Phi(X)^\dagger \, [\, i \partial_\tau - H \,] \, \Phi(X)$ where the expression for the Hamiltonian is just L_o , where we add in the contribution of the two ghost particles in the Hamiltonian. Similarly, we can convert *any* expression in the path integral into operator language via these techniques.

We see, therefore, that the Fadeev-Popov term contributes exactly two degrees of freedom to the first quantized action, which is precisely the number that we need to get the partition function to the 24th power. In the light cone formulation, the partition function naturally emerges to the 24th power because we have explicilty removed the + and - modes of the string. However, in the conformal gauge, the 24th power emerges because the two fermionic variables contribute powers of -2 to the partition function.

The transition to the second quantized formalism is now slightly modified because the fermionic variable contains a τ derivative. Hence, when we begin τ-slicing the path integral we must sum over a complete set of fermionic bra and ket vectors. This naturally leads to a field theory with the following structure:

$$L = \Phi^\dagger(X,\theta)\,[L_0-1]\,\Phi(X,\theta)\,;\quad \Phi(X,\theta) = \langle\Phi|X,\theta\rangle$$

where L_0 contains a sum over both bosonic and ghost fermionic variables.

This action is manifestly Lorentz covariant, but lacks any gauge symmetry. Recently, new developments in BRST quantization of the covariant string [17] have made it possible to re-write the previous Lorentz covariant action in a form which possesses BRST invariance. Kato and Ogawa constructed the integrated BRST current, written as

$$Q_{BRST} = \int_0^\pi d\sigma\, J_{BRST}\,;\quad Q^2 = 0$$

They proved that the charge was nilpotent only in 26 dimensions with zero intercept (which is to be expected if the formalism is to be self-consistent, since this is the result of the conformal anomaly). Siegel [18] then abstracted from this a BRST field theory action:

$$L = \delta_Q[\int d\theta_0\ \Psi\theta_0\bar{\theta}_0\ \Psi]$$

Because the BRST charge Q is nilpotent, the above action is manifestly invariant under the transformation $\delta\Psi = Q\,\Psi$. Because the BRST theory includes the

17. M. Kato and K. Ogawa, Nucl. Phys. B212 (1983) 443.
18. W. Siegel, Phys. Lett. 149B (1985) 157, 162.

original gauge fields plus many other fields, this formulation has been useful in extracting out the gauge covariant field theory in a local fashion. The only problem, then, is to truncate the unwanted fields in the action while maintaining gauge invariance.

5. Gauge Covariant String Field Theory

We now wish to generalize our understanding of string field theory by proceeding to the gauge covariant case.

The lesson we derive from Yang Mills theory and gravity is that, at the free level, the variation of the fields serves to eliminate the unwanted longitudinal states of the theory. For the string, the unwanted longitudinal states are represented by $L_{-n} \mid R >$ so as a guess it is natural to assume that the free gauge principle is represented by:

$$\delta\Phi = \sum_n L_{-n} \Lambda_n$$

In other words, the rotation of an arbitrary field functional is given by the set of longitundinal states of the theory. We can extract from this the variation of the various particles in the spectrum, and we find that $\delta A^a{}_\mu = \partial_\mu \Lambda^a$ for the Yang Mills field, and $\delta h_{\mu\nu} = \partial_\mu \Lambda_\nu + \partial_\nu \Lambda_\mu$ for the graviton field.

In order to extract the action which incorporates this principle, we will find it useful to use the same trick as before, starting with the Feynman path integral and making the transition to the second quantized action. We must be cautious, however, because the Green's functions are not well-defined until we make an explicit choice of the gauge.

We begin as before, by starting with the path integral and then carefully slicing up the path of the string into infinitesimal τ slices.

To handle the infinite vibratory modes of the string, it will be convenient to introduce harmonic oscillators:

$$\mid X_k > \equiv \prod_{k=1}^{\infty} \mid X_{k,n} > \equiv \prod_{k=1}^{\infty} \exp\left(-\tfrac{1}{2}X_{n,k}^2 - i\sqrt{2}X_{n,k}\, a^\dagger{}_n + \tfrac{1}{2}a^\dagger{}_n\, a^\dagger{}_n\right) \mid 0>$$

$$\mid P_k > \equiv \prod_{n=1}^{\infty} \mid P_{n,k} > \equiv \prod_{n=1}^{\infty} \exp\left(-\tfrac{1}{2}P_{n,k} + \sqrt{2}P_{n,k}\, a^\dagger{}_n - \tfrac{1}{2}a^\dagger{}_n\, a^\dagger{}_n\right) \mid 0>$$

$$[a_n, a^\dagger{}_m] = \delta_{mn}\,; \quad < P \mid X > = \prod_{n=1}^{\infty} 2^{-1/2} \exp - i P_n X_n\,; \quad < P \mid P' > = \prod_{n=1}^{\infty} \sqrt{\pi}\,\delta(P_n - P'_n)$$

$$<X|X'> = \prod_{n=1}^{\infty} \delta(X_n - X'_n)$$

$$1 = \int \prod_{n=1}^{\infty} \pi^{-1/2} |P_n>dP_n<P_n|$$

$$= \int \prod_{n=1}^{\infty} |X_n>dX_n<X_n|$$

$$= \prod_{n=1}^{\infty} \sum_{q=1}^{\infty} a^{\dagger q}_n |0><0| a^q_n (q!)^{-1}$$

$$P_n |P> = 2^{-1/2}(a_n + a^\dagger_n)|P$$

$$X_n |X> = i\, 2^{-1/2}(a_n - a^\dagger_n)|X>$$

$$a_n |X> = -i\, 2^{-1/2}(X_n - \partial_{x_n})|X>; \quad a^\dagger_n |X> = i\, 2^{-1/2}(X_n + \partial_{x_n})|X>$$

$$a_n |P> = 2^{-1/2}(P_n - \partial_{P_n})|P>; \quad a^\dagger_n |P> = 2^{-1/2}(P_n + \partial_{P_n})|P>$$

To simplify these expressions, notice that we have explicitly deleted the space-time index μ . To prove these relations, it is convenient to power expand these expressions in terms of Hermite polynomials:

$$H_q(x) \equiv (-1)^q \exp x^2 \partial_x{}^q \exp -x^2 = \exp \tfrac{1}{2}x^2 (x - \partial_x)^q \exp -\tfrac{1}{2}x^2$$

$$<0|a_n{}^q |X> = [\frac{-i}{\sqrt{2}}]^q (X_n - \partial_{x_n})^q <0|X> = [\frac{-i}{\sqrt{2}}]^q \exp \tfrac{1}{2}x_n^2\, H_q(X_n)$$

$$\sqrt{\pi}\,\delta(P_n - P'_n) = <P_n|P'_n> = \sum_{q=1}^{\infty} <P_n|a^{\dagger q}_n|0><0|a^q_n|P'_n>(q!)^{-1}$$

$$= \sum_{q=1}^{\infty} \pi^{-1/2} 2^{-q} (q!)^{-1} H_q(P_n) \exp -\tfrac{1}{2}P_n^2\, H_q(P'_n) \exp -\tfrac{1}{2}P'_n$$

$$\exp iP_n X_n = \sum_{q=1}^{\infty} i^q\, 2^{-q+1/2} (q!)^{-1} H_q(P_n) \exp -\tfrac{1}{2}P_n^2\, H_q(X_n) \exp -\tfrac{1}{2}X_n^2$$

Sandwiched between the various slices, of course we must insert $\delta(P^2 + X'^2)\, \delta(PX')$, which represent the first class constraints taken at each slice. Now let us find an explicit operator expansion of $\Phi(X)$ itself. Let us define $|\Phi>$ as an arbitrary state vector in the Hilbert space of harmonic oscillators. Obviously, this means

$$|\Phi> \equiv \phi(x_o)|0> + A(x_o)^\mu a^\dagger_{1\mu}|0> + g^{\mu\nu}(x_o)\, a^\dagger_{1\mu} a^\dagger_{1\mu} + \cdots$$

where the state $|\Phi>$ contains the tachyon, the photon, a massive graviton, etc. (In ref. [2], we also introduced the vacuum state for the individual quantum fields denoted by $|0\gg$. To be precise, the state $|\Phi>$ must be power expanded around a vacuum state which is the product of $|0>$ and $|0\gg$).

If we now take the matrix element of $|X>$ with $<\Phi|$, then we arrive at the usual expansion of the field Φ in terms of Hermite polynomials [2]:

$$\Phi(X) \equiv <\Phi|X> \equiv \phi(x_o) \prod_{\substack{n=1 \\ \alpha=1,26}}^{\infty} H_o(X_{n,\alpha}) + A^{\mu}(x_o) H_1(X_{1\mu}) \prod_{\substack{n=1 \\ n\neq 1,\mu\neq\alpha}}^{\infty} H_o(X_n) + \cdots$$

Notice that the definition $\Phi(X) \equiv <\Phi|X>$ allows us to make a rigorous transformation from the first quantized formalism to the second quantized formalism starting from Feynman's original path integral expression. There is no more guess work or intuition involved in going from the first quantized formalism to the second quantized field theory.

At each time slice labelled by i, the local gauge invariance guarantee that we have enough symmetry to eliminate the ghost states. In Dirac quantization, we can apply the first class constraints directly onto the state vectors:

$$\int d\sigma \exp in\sigma \left[\hat{P}_{\mu}{}^2 + 2\hat{P}_{\mu} X^{\mu\prime} + X^{\prime 2} \right] |\Phi> \equiv L_n |\Phi> = 0.$$

This means that we are allowed to extract out all the spurious states out of $|\Phi>$ with a projection operator that we will call **P**. We will now give several formulations of this operator **P**:

(a) using the methods pioneered by Brower and Thorn, in writing down a Hilbert space projection operator [19]

$\mathbf{P} = \sum |R_i><R_i|$, which is summed over all real states which satisfy $L_n |R_i> = 0$

(b) using the techiques pioneered by Feigin and Fuchs, in writing down a power expansion over Virasoro operators [20] composed of arbitrarily large products over the elements of the L's $\mathbf{P} = \sum L_{-n} F_n(L_o) L_n + \ldots$

(c) in terms of string operators by extracting the projection operator directly from the Feynman functional at each τ slice.

19. See R.C. Brower and C.B. Thorn, Nucl. Phys. **B31** (1971) 163-182 where the on-shell projection operator is constructed. I am informed by Thorn that they also considered the off-shell case as well, and that they originally introduced the concept of the Shapovalov matrix in their work (private communication).

20. B. L. Feigin and D. B. Fuchs, Sov. Math. Doklady 27 (1983), No. 2, 465-9

It will be helpful to first define some mathematical concepts from the theory of affine Lie (Kac-Moody) algebras [21]: let the *highest weight vacuum vector* be a state $|R>$ which satisfies $L_n\,|R>=0$ for all positive n. We know that we can reproduce the entire representation associated with $|R>$ if we apply all possible lowering step operators. Similarly, let the *universal enveloping algebra* $U(L_{-n})$ consist of all possible products that we can form from L_{-n}. A typical element of the enveloping algebra is $\mathbf{L}_{-I} \equiv L_{-1}^{\lambda_1} L_{-2}^{\lambda_2} \ldots L_{-n}^{\lambda_n} \in U(L_{-n})$

This construction is a simple generalization of familiar techniques found in the theory of finite Lie groups. For example, for O(3), the highest weight vacuum vector is simply $|lm>$ where $m=l$. Notice that the application of a raising operator on this state yields zero: $L_+|ll>=0$. Now, by applying L_- onto this state, we can extract all members of the multiplet of states (singles, triplets, etc.) The universal eveloping algebra for O(3) only consists of the lowering operator L_-

Now let p(N) equal the number of partitions that we can make of the integer N. Let I equal any integer from 1 to p(N). If we let $N=|I|=\sum_{i=1}^{n} i\,\lambda_i$, then I simply counts all possible partitions of the integer $N=|I|$. Now, the set of all states created by applying the universal enveloping algebra (the set of all products of lowering operators) to the hightest weight vacuum vector $|R>$ merely creates the entire multiplet, called the Verma module: $V(h,c)=U(L_{-n})|R>$ where $L_0\,|R>=h\,|R>$ and c is the dimension of space time appearing in the universal extension (Schwinger term). Notice that all the vectors in the multiplet or the Verma module are spurious states except the highest weight vacuum vector $|R>$. Using the notation of Feigin-Fuchs, let us now define $|I> \equiv \mathbf{L}_{-I}\,|0>$. Then we can define the Shapovalov matrix as [22]:

$$S_{IJ}{}^{N} = <I\,|J> = <0|\mathbf{L}_I\;\mathbf{L}_{-J}\,|0>$$

for $|I|=|J|=N$ and $h=p^2$. *Given the Shapovalov matrix, we can now write down an expression for* **P** *to all orders.* Let us define $\mathbf{P}^{R_n}$ as the projection operator which projects onto real states at the nth level, while $\mathbf{P}^{S_n}$ is the off mass shell spurious particle projection projection operator at that level. For example,

$$\mathbf{P}^{R_0}=|0><0|$$

21. V. Kac, *Infinite Dimensional Lie Algebra* Birkhauser, Boston, 1983.

$$\mathbf{P}^{R_1} = 1_1 - P^{S_1}; \quad \mathbf{P}^{S_1} = (2p^2)^{-1} L_{-1} \mathbf{P}^{R_0} L_1$$

$$P^{R_2} = 1_2 - \mathbf{P}^{S_2}$$

$$\mathbf{P}^{S_2} = [4p^2 + D/2] | S^{-1} | L_{-1}{}^2 \mathbf{P}^{R_0} L_1{}^2 - 6p^2 | S^{-1} | [L_{-1}{}^2 \mathbf{P}^{R_0} L_2 + L_{-2} \mathbf{P}^{R_0} L_1{}^2]$$

$$+ 4p^2(1+2p^2) | S^{-1} | [L_{-2} \mathbf{P}^{R_0} L_2] + (2+2p^2)^{-1} L_{-1} \mathbf{P}^{R_1} L_1$$

$$= \sum_{I,J=1}^{2} | I >_2 (S^{-1})_{IJ}{}^2 (p^2) < J |_2 + (2+2p^2) L_{-1} \mathbf{P}^{R_1} L_1$$

$$|1>_2 \equiv L_{-1}{}^2 |0>; \quad |2>_2 \equiv L_{-2} |0>$$

$$S_{IJ}{}^2 = \begin{vmatrix} 8p^4+4p^2 & 6p^2 \\ 6p^2 & 4p^2+D/2 \end{vmatrix} = < I |_2 | J >_2$$

$$\mathbf{P}^{R_3} = 1_3 - \mathbf{P}^{S_3}$$

$$= \sum_{I,J=1}^{3} | I >_3 (S^{-1})_{IJ} (p^2) < J |_3 + \sum_{IJ=1}^{2} \mathbf{L}_{-I} \mathbf{P}^{R_1} (S^{-1})_{IJ}{}^2 (p^2+1) + (2p^2+4) L_{-1} \mathbf{P}^{R_2} L_{-1}$$

$$|1>_3 \equiv L_{-1}{}^3 |0>; \quad |2>_3 \equiv L_{-1} L_{-2} |0>; \quad |3>_3 \equiv L_{-3} |0>$$

$$S_{IJ}{}^3 = \begin{vmatrix} 24p^2(1+p^2)(1+2p^2) & 36(p^2+p^4) & 24p^2 \\ 36(p^2+p^4) & 8p^4+(34+D)p^2+2 & 16p^2+2D \\ 24p^2 & 16p^2+2D & 6p^2+2D \end{vmatrix} = < J |_3 | J >_3$$

$$| S_{IJ}{}^3 | = 24p^2 | S_{IJ}{}^2 | [3p^4 + (D-7)p^2 + D + 2]$$

By examining the iteration in lowest orders, we can now easily generalize our results to the arbitrary case: calculating the nth level from the n-1 level (levels are ranked by the eigenvalue of the number operator on the Hilbert space):

$$\mathbf{P} = \sum_{N=0}^{\infty} \mathbf{P}^{R_N} = \sum_{N=0}^{\infty} 1_N - \mathbf{P}^{S_N}$$

$$\mathbf{P}^{S_N} = \sum_{M=1}^{N} \sum_{\substack{I,J=1 \\ |I|=|J|=M}}^{p(N-M)} \mathbf{L}_{-I} (S^{-1})_{IJ}^{N-M} (p^2+M) \mathbf{P}^{R_M} \mathbf{L}_J$$

It is easy to show that $\mathbf{P}^2 = \mathbf{P}$, $\mathbf{P} | R > = | R >$. $< R | \mathbf{P} = < R |$, and that $< R | L_n \mathbf{P} = \mathbf{P} L_{-n} | R > = 0$ for any highest weight vacuum vector $|R>$. *Notice that if the Shapovalov matrix known, then we know* $\mathbf{P}$ *to all orders.* Since the divergences of the Shapovalov are well-known and have been isolated by Kac [23], we know that it is possible over a wide range of parameters p^2 and c to

23. V. Kac, Lect. Notes in Physics, vol. 94, 441-5 (1979)

construct the *off-shell* projection operator.

It is easy now to see that, at least for the free field theory, this reproduces the gauge invariance of the Maxwell field and the graviton field (when we generalize to closed strings). This is because $L_{-1} \approx k^{\mu} a^{\dagger}{}_{1\mu}$. When acting on an arbitrary string functional $\Phi(X)$:

$$\delta|\Phi> = \delta A^{\mu} a^{\dagger}{}_{1\mu}|0> + \cdots = L_{-1}|\Lambda> + \cdots = \delta L_{-1}\Lambda(x_o)|0> + \cdots$$

This implies that $\delta A_{\mu} = \partial_{\mu}\Lambda$ for the open string and $\delta h_{\mu\nu} = \partial_{\mu}\Lambda_{\nu} + \partial_{\nu}\Lambda_{\mu}$ for the closed string case.

Rather than power expand around projection operaters defined at each level, yet another way in which to derive the total projection operator is with a power expansion in the universal enveloping algebra $U(L_{-n})$

Rather than expand $\mathbf{P}$ in terms of $\sum|R_i><R_i|$, we can also expand in all orders in elements of the universal enveloping algebra and its conjugate. Following the method of Feigin-Fuchs, who created a generalized Casimir operator by this same power expansion, we can show that $\mathbf{P} = 1 - \mathbf{P}^S$ and

$$\mathbf{P} = 1 + -L_{-1}(2L_o)^{-1}L - L_{-2}(4L_o+2)^{-1}\Delta^{-1}L_2 + 3L_{-2}\Delta^{-1}L_1{}^2$$

$$+ 3L_1{}^2\Delta^{-1}L_{-2} + L_{-1}{}^2 \Big| [4(L_o{}^2+1]^{-1} - (4L_o + D/2)(2L_o\Delta)^{-1} \Big| L_1^2 + \cdots$$

This easily generalizes to:

$$\mathbf{P}^S = \sum_{N=1}^{\infty} \sum_{\substack{I\,J=1 \\ |I|=|J|=N}}^{p(N} \mathbf{L}_{-I}\; F_{IJ}^N(L_o)\; \mathbf{L}_J$$

$$F_{IJ}^N(L_o) \equiv \Big[\delta_{IL} - A_{IL}^{N-1}(L_o)\Big] (S^{-1})_{LJ}^N(L_o$$

$$\sum_{M=1}^{N-1} \sum_{\substack{I\,J=1 \\ |I|=|J|=N}}^{p(M)} \mathbf{L}_{-I}\; F_{IJ}^M(L_o)\,\mathbf{L}_J\;\; \mathbf{L}_{-S}|R> \equiv \mathbf{L}_{-P}\;\; A_{PS}^{N-1}(L_o)|R>$$

where $S_{IJ}^N(L_o)|R> \equiv L_I L_{-J}|R>$ and S^N is determined iteratively from A^{N-1} . Notice that we have, to all orders, an expression for the projection operator in terms of a simple power expansion in operators composed out of the universal enveloping algebra and the Shapovalov matrix.

6. Field Theory of Super Strings

These methods obviously can be generalized to the superstring and the heterotic string case. Using the same methods as before, we find easily: $G_{-I} \equiv G_{-1/2}^{\lambda_{-1/2}} G_{-3/2}^{\lambda_{3/2}} \cdots G_{-n}^{\lambda_n}$ and $S_{IJ} \equiv \langle 0 | G_I G_{-J} | 0 \rangle$ where

$$[G_m, G_n]_+ = 2L_{m+n} + \tfrac{1}{2}D(m^2 - 1/4)\delta_{m,-n}$$

$$[L_m, G_n] = (\tfrac{1}{2}m - n)G_{m+n}$$

$$\mathbf{P}^{R_0} = |0\rangle\langle 0|$$

$$\mathbf{p}^{R_{1/2}} = 1_{1/2} - \mathbf{P}^{S_{1/2}}$$

$$\mathbf{P}^{S_{1/2}} = (2p^2)^{-1} G_{-1/2}|0\rangle\langle 0| G_{1/2}$$

$$\mathbf{P}^{R_1} = 1_1 - \mathbf{P}^{S_1}$$

$$\mathbf{P}^{S_1} = (2P^2)^{-1} L_{-1}|0\rangle\langle 0| L_1 + (2p^2+1)^{-1} G_{-1/2}\mathbf{P}^{R_{1/2}} G_{1/2}$$

$$\mathbf{P}^{R_{3/2}} = 1_{3/2} - \mathbf{P}^{S_{3/2}}$$

$$\mathbf{P}^{S_{3/2}} = (4p^2+2)|S^{-1}| G_{-3/2}|0\rangle\langle 0| G_{3/2} + (2p^2+D)|S^{-1}| G_{-1/2}L_{-1}|0\rangle\langle 0| L_1 G_{1/2}$$
$$+ (-4p^2)|S^{-1}| \left[G_{-3/2}|0\rangle\langle 0| L_1 G_{1/2} + G_{-1/2}L_{-1}|0\rangle\langle 0| G_{-3/2} \right]$$
$$+ (2p^2+1)^{-1} L_{-1}\mathbf{P}^{R_{1/2}} L_1 + G_{-1/2}\mathbf{P}^{R_1} G_{1/2}$$

We can now easily generalize our result to the arbitrary case:

$$\mathbf{P} = \sum_{N=0}^{\infty} \left\{ 1_N - \sum_{M=1/2}^{N} \sum_{\substack{I\,J=1/2 \\ |I|=|J|=M}}^{p(N-M)} G_{-I} (S^{-1})_{IJ}^{N-M}(p^2+M)\, \mathbf{P}^{R_M}\, G_J \right\}$$

Notice that the generalization to the Neveu-Schwarz model is straightforward, since G is the square root of L (p represents the number of distinct partitions of a half-integer).

The same methods apply to the Ramond model. The algebra in this case is given by:

$$[L_m, F_n]_- = (\tfrac{1}{2}m - n) F_{n+m}$$

$$[F_m, F_n]_+ = 2L_{m+n} + \tfrac{1}{2}Dm^2\delta_{m,-n}$$

$$[L_m, L_n]_- = (m-n)L_{m+n} + (D/8)m^3\delta_{m,-n}$$

Using the same methods as before, the first few terms in the projection operator are given by:

$$\mathbf{P}^{R_o} = |0\rangle_F \langle 0|_F$$

$$\mathbf{P}^{R_1} = 1_1 - \mathbf{P}^{S_1}$$

Using identical techniques, this can obviously be generalized to the arbitrary case.

With these definitions, we can now write the Neveu-Schwarz-Ramond model in a concise fashion:

$$L = \Phi(X,\Gamma)^\dagger \mathbf{P}\left[i\partial_\tau - (L_o - 1)\right]\mathbf{P}\,\Phi(X,\Gamma) + \overline{\Psi}(X,\Gamma)\mathbf{P}\left[i\partial_\tau - (F_o - \tfrac{1}{2})\right]\mathbf{P}\,\Phi(X,\Gamma)$$

Notice that this action can be easily modified so that the sum only contains even G parity states. Written in this elegant fashion, we see that it is an obvious generalization to assume that the above action is invariant under the supersymmetric variation:

$$\delta\Phi = A\Psi\,;\ \delta\Psi = B\Phi$$

where A and B are related to the strange matrices found earlier when studing the fermion vertex function which interchanges the Hilbert space of b operators for the Hilbert space of d operators. (These matrices A and B have been written down explictly for the light cone case, but only have been constructed for the lower lying states for the covariant case.) This shows that the Neveu-Schwarz-Ramond model can (at least

in principle) be written in a fashion which is manifestly supersymmetric. This is quite ironic, because supersymmetry (at least in 2 dimensions) was first discovered [24] explicitly in the string model and later generalized to four

24. J.L. Gervais and B. Sakita, Nucl. Phys. B34 (1971) 477.

dimensions, but it was only years later that it was suspected that the Neveu-Schwarz-Ramond model itself was invariant under a space-time supersymmetric transformation [25]. In our second quantized formalism, however, it is relatively easy to show that the Neveu-Schwarz-Ramond model is supersymmetric (if we can form the matrices A and B).

The Neveu-Scharz-Ramond model, we see, is (formally) supersymmetric at the second quantized level in 10 dimensions but not at the first quantized level. The Green-Schwarz action, however, is much better from the point of view of supersymmetry, since it is manifestly supersymmetric in 10 dimensions even at the first quantized level. However, the Green-Schwarz action suffers from complications when we quantize it covariantly. We begin with the first quantized action of Green and Schwarz:

$$L = -\sqrt{-g} - i\,\partial_\tau X_\mu\,\bar{\theta}\gamma^\mu\partial_\sigma\rho\theta + i\,\partial_\sigma X_\mu\,\bar{\theta}\gamma^\mu\partial_\tau\rho\theta + \bar{\theta}^1\gamma_\mu\partial_\tau\theta^1\,\bar{\theta}^2\gamma^\mu\partial_\sigma\theta^2 - \bar{\theta}^1\gamma_\mu\partial_\sigma\theta^1\,\bar{\theta}^2\gamma^\mu\partial_\tau\theta^2$$

where $g = -(\Pi_\tau\Pi_\sigma)^2+(\Pi_\tau\Pi_\tau)(\Pi_\sigma\Pi_\sigma)$ and

$$\Pi_{\tau\mu}=\partial_\tau X_\mu-i\,\bar{\theta}\gamma_\mu\partial_\tau\theta; \quad \Pi_{\sigma\mu}=\partial_\sigma X_\mu-i\,\bar{\theta}\gamma_\mu\partial_\sigma\theta$$

In order to make the transition from a classical theory to a quantum theory, we will find it covenient to reformulate the model in terms of the second order formalism, where we introduce canonical conjugates P_μ and $\bar{p}_\theta$ and double the number of independent fields. It is straightforward to write down their action in second order formalism, using the conjugate pairs (P,X) and ($\bar{p}^a_\theta$, θ^a) . θ is a Majorana-Weyl spinor in 10 dimensional space with 16+16 independent components. Notice that we are explicitly introducing the spinor conjugate to the spinor. If we define $P_\mu\equiv\frac{\delta L}{\delta \dot{X}^\mu}$ and $\bar{p}_\theta\equiv\frac{\delta L}{\delta\dot{\theta}}$ we find a new expression for the Green-Schwarz action in second order form. After a certain amount of algebra, we find that their theory can be *exactly* expressed as:

$$Z = \int DX\; DP\; D\theta\; D\bar{p}_\theta\; D\lambda\; D\rho\; D\epsilon^a\; \Delta \exp i \int L(P,X;\, \bar{p}^a_\theta, \theta^a)\, d\sigma\, d\tau$$

$$L(P,X;\bar{p}^a_\theta,\theta^a) = P_\mu\dot{X}^\mu - \bar{p}_\theta\dot{\theta} + \frac{\lambda}{2}(\pi_o^2+\pi_1^2) + \rho(\pi_o\pi_1) + \bar{Q}^a\epsilon^a$$

$$\pi_o^2 + \pi_1^2 \equiv 0 \quad ; \quad \pi_o\pi_1 \equiv 0$$

$$\pi_{o\mu} \equiv P_\mu+\bar{\theta}\gamma_\mu\partial_\sigma\rho\theta \quad ; \quad \pi_{1\mu} \equiv \partial_\sigma X_\mu - i\,\bar{\theta}\gamma_\mu\partial_\sigma\theta$$

---

25. F. Gliozzi, J. Scherk, P. Olive, Nucl. Phys. B122 (1977) 253.

$$\bar{Q}^1 \equiv \bar{p}_\theta^1 - i\bar{\theta}^1\gamma_\mu\pi_o^\mu + i\bar{\theta}^1\gamma_\mu\pi_1^\mu - \bar{\theta}^1\gamma_\mu\partial_\sigma\theta^1\,\bar{\theta}^1\gamma^\mu \equiv 0$$

$$\bar{Q}^2 \equiv \bar{p}_\theta^2 - i\bar{\theta}^2\gamma_\mu\pi_o^\mu - i\bar{\theta}^2\gamma_\mu\pi_1^\mu + \bar{\theta}_2\gamma_\mu\partial_\sigma\theta^2\,\bar{\theta}^2\gamma^\mu \equiv 0$$

where ρ^{ab} is the Pauli matrix σ_z and λ,ρ,ϵ are Lagrange multipliers

Taking linear combinations of these constraints, we can easily diagonalize our primary constraints so that the fermionic and bosonic components decouple:

$$L_+ \equiv \pi_o^2 + \pi_1^2 - 2\bar{Q}\rho\partial_\sigma\theta \equiv P^2 + X'^2 + 2\bar{p}_\theta\rho\partial_\sigma\theta \equiv 0.$$

$$L_- \equiv 2\pi_o\pi_1 - 2\bar{Q}\partial_\sigma\theta \equiv 2PX' - 2\bar{p}_\theta\partial_\sigma\theta \equiv 0.$$

Notice that the fermionic and bosonic fields have now completely decoupled from each other in L_o . We now proceed to the quantization of the model. Whenever the momenta appear in the path integral, they can be replaced by a partial derivative because:

$$P_\mu \exp iP\dot{X} = -i\,\partial_\mu \exp iP\dot{X}$$

$$\bar{p}_\theta \exp -i\bar{p}_\theta\dot{\theta} = -i\,\partial_\theta \exp -i\bar{p}_\theta\dot{\theta}$$

Using this replacement of c-numbers by their operator counterpart, we easily find the canonical commutation relations:

$$\Big[P_\mu(\sigma), X_\nu(\bar{\sigma})\Big]_- = -i\,\delta_{\mu\nu}\,\delta(\sigma-\bar{\sigma}); \qquad \Big\{\bar{p}_\theta^\alpha(\sigma), \theta^\beta(\bar{\sigma})\Big\}_+ = -i\,\zeta^{\alpha\beta}\,\delta(\sigma-\bar{\sigma})$$

$$\int d\sigma\, d\bar{\sigma} \Big\{\bar{Q}(\sigma)\epsilon^a(\sigma), \bar{Q}(\bar{\sigma})\epsilon^b(\bar{\sigma})\Big\}_- = \int d\sigma\,[\,\bar{\epsilon}^{[a}\gamma_\mu(\pi_o^\mu - \pi_1^\mu\rho)\,\epsilon^{b]}\,]$$

$$\int d\sigma\, d\bar{\sigma}\Big[f(\sigma)L_\pm(\sigma), g(\bar{\sigma})L_\pm(\bar{\sigma})\Big]_- = -2i\int d\sigma\,(f'g - g'f)L_-(\sigma)$$

$$\int d\sigma\, d\bar{\sigma}\Big[f(\sigma)L_+(\sigma), g(\bar{\sigma})L_-(\bar{\sigma})\Big]_- = -2i\int d\sigma\,(f'g - g'f)L_+(\sigma)$$

where ζ equals the projection onto Majorana-Weyl spinors.

Let us now examine the counting of fermion states. Formally, it appears as if we have 16+16 first class constraints in the Q's. But this cannot be true, because local supersymmetry only has 8+8 components. Therefore, *Q is a non-trivial mixture of first and second class constraints*. This is the origin of the problem found earlier [26] in attempts to quantize the superstring. A naive quantization of the theory will have terms proportional to $1/\pi^2$, which is not well-

---

26. M.B. Green and J.H. Schwarz, Nucl. Phys. **B243** (1984) 285-306

defined. This problem arises because the original action contained terms proportional to $\bar{\theta}\pi_\mu\gamma^\mu\dot{\theta}$. A naive quantization of the model would define the canonical conjugate to θ as $\bar{\theta}\pi_\mu\gamma^\mu$, which would result in the commutator between two θ 's having a term proportional to $1/\ \pi_\mu\gamma^\mu$, which is absurd.

Notice that the algebra of the Q's do not close on the π's , a sure sign that there are second class constraints in the theory. The problem is that it is formally impossible to decompose a 16 dimensional Majorana-Weyl spinor down to 8 components. The 16 component spinor is the smallest spinor representation of the Lorentz group.

In the light cone gauge, this problem is simply solved by contracting the spinor onto $k_\mu\gamma^\mu$ where the fact that $k^2=0$ forces the spinor $k_\mu\gamma^\mu\theta$ to have only eight independent components, i.e. it satisfies eight linear constraints $k_\mu\gamma^\mu(k_\nu\gamma^\nu\theta) = 0$. Using this analogy, we realize that we actually have a non-zero vector in the theory, the vector $\pi_{\pm\mu}$. Using this fact, we can now construct projection operators for the Green-Schwarz theory which hold because of the primary constraint $\pi^2=0$:

$$\mathrm{P}_a = 1 + \gamma^\mu\gamma^\nu\ [\pi_{\mu-}\pi_{\mu+}]/\ [2\pi_+\pi_-]; \quad \mathrm{P}_b = 1 + \gamma^\mu\gamma^\nu[\pi_{\mu+}\pi_{\mu-}]/\ [2\pi_+\pi_-]$$

where $\pi_\pm \equiv \pi_o \pm \pi_1$, and $\mathrm{P}_a + \mathrm{P}_b = 1;\quad \mathrm{P}_a\mathrm{P}_b = \delta_{ab}\mathrm{P}_a$

$$\pi_{\mu+}\gamma^\mu\mathrm{P}_a = \pi_{\mu-}\gamma^\mu\mathrm{P}_b = \mathrm{P}_a\pi_{\mu-}\gamma^\mu = \mathrm{P}_b\pi_{\mu+}\gamma^\mu = 0.$$

Notice that these identities hold modulo the first class constraints. We can now reduce 16+16 component Majorana-Weyl spinors down to 8+8 component spinors without violating Lorentz invariance. Let us, therefore, write down the new set of 8+8 first class spinor constraints which generate supersymmetry:

$$\bar{S}^1 \equiv \bar{Q}^1\ \pi_{\mu-}\gamma^\mu; \quad \bar{S}^2 \equiv \bar{Q}^2\ \pi_{\mu+}\gamma^\mu \tag{6.8}$$

and the set of 8+8 second class constraints:

$$\bar{T}^1 \equiv \bar{Q}^1\ \pi_{\mu+}\gamma^\mu; \quad \bar{T}_2 \equiv \bar{Q}^2\ \pi_{\mu-}\gamma^\mu \tag{6.9}$$

We will find that the first class constraints do indeed close, while the second class constraints do not close, as required:

$$\int d\sigma\, d\bar{\sigma} \left| f(\sigma)L_\pm(\sigma), \bar{S}^a(\bar{\sigma})\,\epsilon(\bar{\sigma}) \right|_- = (-1)^{\frac{1}{2}(a\pm a)}\, i \int d\sigma\, \bar{S}^a(f'\epsilon - f\epsilon')$$

$$\int d\sigma\, d\bar{\sigma} \left| \bar{S}^a(\sigma)\epsilon(\sigma), \bar{S}^b(\bar{\sigma})\eta(\bar{\sigma}) \right|_- =$$

$$\delta^{ab}\int d\sigma \left| \bar{\epsilon}\pi^a_\mu\gamma^\mu\eta\ \ (L_+ + (-1)^a L_-) + (-1)^a[8\bar{S}^a\eta\,\bar{\epsilon}\partial_\sigma\theta^a + 2\bar{S}^a\gamma^\mu\partial_\sigma\theta^b\ \ \bar{\eta}\gamma_\mu\epsilon \right| - (\epsilon \longleftrightarrow \eta)$$

$$\int d\sigma d\bar{\sigma} \left[\bar{T}^a(\sigma)\epsilon(\sigma), \bar{T}^b(\bar{\sigma})\epsilon(\bar{\sigma}) \right]_- = -2\delta^{ab} \int d\sigma(\pi_o^2 - \pi_1^2)\bar{\epsilon}\pi_\mu^a \gamma^\mu \eta$$

where $\pi^1 = \pi_-$; $\pi^2 = \pi_+$ and modulo terms which vanish on the constraints. However, we cannot as yet make the transition to the second quantized field theory from the Feynman path integral. The reason is because second class constraints cannot naively be applied on the state vectors of the theory. Because the second class constraints do not close on themselves, the application of the anti-commutator of two second class constraints onto state vectors will lead to contradictions. Because of this, we will be forced to adopt the formalism of Dirac brackets. Functionally, this simply means that there is an extra delta function in the measure corresponding to the 8+8 second class constraints. In order to proceed to the second quantized theory, we must construct the projection operator onto real states out of the Dirac brackets. This means that we must adopt the following modification of all naive commutators:

$$[X, Y] \rightarrow [X, Y]^* = [X, Y] - [X, \phi^a](C^{-1})^{ab}[\phi^b, Y]$$

where ϕ represents the second class constraints given by the T's and the C matrix is simply proportional to the inverse of $\pi_+\pi_-$. Notice that this operator can be power expanded into normal modes, $\int d\sigma \exp in\sigma [\pi_+\pi_-]^{-1}$, much like the usual Virasoro generators, because the inverse operator is well-defined, unlike the operator $\pi_\pm^2$, which is proportional to the first class constraints and hence has no inverse. Unfortunately, however, this considerably complicates normal ordering of the operators.

At this point, notice that the theory is completely well-defined. All commutators now close among themselves because all constraints are now first class. We can now construct the Verma module corresponding to this theory in exactly the same way as before and write down the second quantized field theory, which has the same action as before except that the number of fields is much larger and we must take into account the Dirac brackets.

7. Field Theory of Heterotic Strings

It is now a straightforward exercise to second quantize the heterotic string of Gross, et. al. [27], the most promising candidate for a field theory of

27. D.J. Gross, J.A. Harvey, E. Martinec, and R. Rohm, Princeton preprint, Jan. and July 1985.

everything. In the right moving sector, we can always put the 10 dimensional Neveu-Schwarz-Ramond model or the Green-Schwarz model (for details, see [7]). For the left-moving sector, given by compactifying 16 of the 26 dimensions describing the bosonic string, we must use the language of Kac-Moody algebras.

We will derive the field theory action for the heterotic string using the same techniques as before (a) putting the first quantized action in second order form (b) using the Feynman path integral to make the transition from first to second quantized form (c) using the formalism of Kac to construct the projection operator.

The non-trivial feature of the left-moving sector is that we must compatify on even self-dual lattices in 16 dimensions, of which there are only two: $E_8 x E_8$ and spin(32) / Z_2. Let us begin with a string X defined only over 16 dimensions:

$$X^I(\theta) = x^I + p^I \theta + i \sum_{n \neq 0} \alpha_n \exp in\theta / n$$

Let us now define the vertex operator

$$q^I(\theta) = c^I : \exp i X^I(\theta) :$$

where c^I defines a "twist" [28]. It is now straightforward to construct operators which will create the various excitations of both the Kaluza-Klein tower as well as the usual Regge tower:

$$J(I,\theta) = :q_I^*(\theta) q_I(\theta): = -\frac{d}{d\theta} X^I(\theta)$$

$$J(\alpha,\theta) = :q_I^*(\theta) q_J(\theta): ; \quad \alpha = e_I - e_J$$

$$L(\theta) = \frac{N}{2} [\sum : J(I,\theta)^2 : + \sum_\alpha : J(\alpha,\theta) J(-\alpha,\theta) :]$$

where α represents the root vectors of the lattice, the e's represent orthonormal vectors, and N = $[1+c_2(G)/2]^{-1}$ and $c_2(G)$ equals the value of the quadratic Casimir operator for the adjoint representation. It is then straightforward to derive the algebra that these operators form:

$$[J(I,\theta), J(K,\theta')] = 0$$

28. We adopt the notation of T. Eguchi and K. Higashijima, Univ. of Tokyo preprint, presented at the Niels Bohr Centennial, Copenhagen, 1985. See also the work of L. Dolan and R. Slansky, Los Alamos preprint LA-UR-85-324.

$$[J(I,\theta), J(\alpha,\theta')] = -2\pi\,\delta(\theta-\theta')\ \alpha_I\, J(\alpha,\theta)$$

$$[J(\alpha,\theta), J(-\alpha,\theta')] = 2\pi\,\delta(\theta-\theta') \sum_I \alpha_I\, J(I,\theta) + 2\pi i\,\delta'(\theta-\theta')$$

$$[J(\alpha,\theta), J(\beta,\theta')] = 2\pi\,\delta(\theta-\theta')\, J(\alpha+\beta) \quad \text{if } \alpha+\beta \in \Lambda;\ =0 \ \textit{otherwise}$$

$$[L(\theta), J(\alpha,\theta')] = J(\alpha,\theta)\delta'(\theta-\theta') \quad \text{if } \alpha \in G$$

$$[L(\theta), L(\theta')] = 2\pi i\,\delta(\theta-\theta')[L(\theta)+L(\theta')] - \frac{c}{12}2\pi i[\,\delta'''(\theta+\theta') + \delta'(\theta-\theta')]$$

where c = n-1 for SU(n); c = n + ½ for SO(2n+1); c = n(2n+1)/(n+2) for SP(n); c = n for SO(2n) and E_n .

With these commutators, one can generate the entire spectrum of states for the left-moving sector of the theory. For example, notice that the J's simply generate the Kaluza-Klein and Regge towers for the theory. But because these J's are part of a larger $E_8 x E_8$ or spin (32) / Z_2 algebra, we can show that the Kaluza-Klein and Regge towers can be re-grouped according to representations of the algebra. It is a simple although painstaking matter, therefore, to create the projection operator for the entire theory.

One distrubing feature of these commutators, however, is that the Virasoro generator L does not commute with the Kac-Moody generators. This means that the total algebra is the *semi-direct* product, not the direct product between the graded Virasoro algebra and the Kac-Moody algebra. Although mathematically beautiful, this is an unpleasant feature of the algebra, because it means that the Kac-Moody generators create ghost or spurious states.

At first, this seems to be a fundamental defect in our construction. From the light cone theory of the heterotic string, we know that the states of the theory are simple direct products of the light cone transverse states with the Kac-Moody states. However, if the commutator between L and J does not vanish, it means that the operators J create ghost states and the counting seems to be fundamentally flawed.

Originally, this was seen to be a serious obstruction to the creation of the field theory of covariant heterotic strings. If we cannot reproduce the counting of states of the light cone theory, then surely there is something wrong with our approach. Upon deeper reflection, however, we find that there is a simple resolution of this puzzle.

The key observation is to realize that the Virasoro operator L is the *sum* of the Virasoro generator coming from the 16 compactified dimensions with the Virasoro generator coming from the 10 uncompactified dimensions. As a consequence, real states are non-trivial mixtures of operators of both sectors, but the total counting of states is the same as the light cone theory. In other words, the operators J, indeed, create ghost states, but this is irrelevant to the counting of real states because real states of the theory are mixtures of the two sectors. Both real states and ghost states are mixtures of the two sectors but the total counting of states remains exactly as in the usual light cone theory.

For example, the Virasoro gauges immediately eliminate the longitudinal state of the spin 1 field, leaving the state

$$a^{\dagger}_{1\mu} \rightarrow a^{\dagger}_{1\mu} - k_\mu k_\alpha a^{\dagger\alpha}_{1} / k^2$$

For the compactified case, the sum over alpha will now include the summation over the compactified indices as well as the usual 10 dimensional ones. Thus, the Kac-Moody operators do, indeed, create ghost states, but the total number of states eliminated by the Virasoro algebra remains exactly as before. We therefore reproduce the usual counting of the light cone states.

Now, let us put everything together. The field functional will now be represented by the direct product of the left and right moving sector:

$$|\Phi\rangle = \sum_{b,k} \phi^{ik} \, |b\rangle_R \, |k\rangle_L$$

$$|\Psi\rangle = \sum_{f,k} \psi^{jk} \, |f\rangle_R \, |k\rangle_L$$

where L and R represent left and right moving sectors, the ϕ and ψ represent bosonic and fermionic fields of arbitrarily high spin, and b (f) represent the complete set of bosonic (fermionic) states for that particular sector.

The complete action for the heterotic string now reads:

$$L = \Phi^\dagger \mathrm{P}_B \, (L_o - \alpha_B) \, \Phi + \overline{\Psi} \mathrm{P}_F \, (L_o - \alpha_F) \, \Psi$$

It should be noted that this action, like all the gauge covariant actions discussed so far, is non-local because of the presence of inverse powers of momenta in the inverse Shapovalov matrix. These non-local terms, of course, disappear completely when we take the gauge $L_n \; |\Phi\rangle = 0$. The non-locality is therefore no problem for the quantization of the model. However, it is possible to add an infinite number of auxiliary fields to absorb this non-locality [29]. At present, it seems that there are fields which will make the theory local. Which is the correct choice? We feel that there will be only one choice of auxiliary fields which will generalize to the interacting case. At the free level, it is still difficult to decide which set of auxiliary fields is the correct one (because certain fields present at the free level may not couple to the real states of the theory at the interacting level) but this ambiguity should be resolved once we have a better understanding of the interactions.

8. Conclusion

In summary, we have been able to write only the free part of the gauge covariant action for superstrings and heterotic strings.

The next step, of course, is to write down the interaction terms, which we will accomplish in the same way as we extracted the field theory in the light cone gauge, i.e. by carefully factorizing the path integral corresponding to the covariant splitting of strings.

But the real question remains, what is the underlying gauge principle and geometry behind the string? One expects that it should be even more beautiful and elegant than general covariance and O(N) gauge invariance, which is contained within it. However, at this point it still remains a challenge to find this symmetry.

It is our belief that this gauge principle will prove decisive in explaining why we live in four dimensions. We have seen that nature has made maximum use of gauge symmetry when exploring the implications of general relativity and Yang Mills theory. Gauge theory, in fact, has proven to be one of the most important mathematical tools of the theorist in the last half century. If gauge principles are so important, especially to a understanding of non-perturbative effects, it is natural to assume that gauge symmetry will continue to play a profound role in the development of the string model. We expect to find many surprises when we have a better understanding of the rich structure behind gauge theories

---

29. M. Kaku, "Locality in the Gauge Covariant Field Theory of Strings," to appear in Phys. Lett. For an elegant treatment of auxiliary fields derived from the BRST formalism, see the work of K. Itoh, T. Kugo, H. Kunitomo, and H. Ooguri, Kyoto preprint; W. Siegel and B. Zwiebach, UCB-PTH-85/30; T. Banks and M. Peskin, SLAC-PUB-3740; See also the work of A. Neveu and P.C. West, CERN-TH.4200/85.

9. Acknowledgments

We are happy to acknowledge many fruitful discussions with B. Sakita on the field theory of strings. We would like to thank the organizers of the Lewes Superstring Workshop, and especially Louis Clavelli, for their hospitality. We also wish to acknowledge partial support from NSF PHY 82-15364.

COMPACTIFIED STRING THEORY

L. Dolan

The Rockefeller University, New York, New York 10021

September 1985

Abstract

The representation theory of affine Kac-Moody Lie algebras is discussed in the context of compactified dual string models. A new quantum number, affinecharge, is identified for the particle multiplets. The light-cone gauge version of the vertex operator construction is defined and shown to be the relevant spectrum generating algebra of compactified strings. These theories are candidates to unify gravity with the other particle interactions and give realistic low energy phenomenology. Their connection with the theory of finite sporadic groups is described.

I. Introduction

Recent results in unified superstring theory [1] have generated a certain amount of enthusiam that some type of string theory may provide us with a mathematically consistent theory of gravity as well as a theory whose low energy phenomenology has several features in common with realistic models.

These features include a gauge group that can describe the SU(3) x SU(2) x U(1) vector mesons, and that has complex representations to describe the families of quarks and leptons. In certain models there is also absence of possible anomaly terms in one loop.

These theories are dynamical in that they allow the calculation of scattering amplitudes in a well defined systematic approximation and do not merely provide a taxonomy of one particle states. What we have therefore, is a theory with some phenomenology which unifies all interactions and admits a perturbative expansion.

A given string theory fixes the particle content uniquely, and it puts an infinite number of spins with different masses and internal quantum numbers in the same multiplet. A classification of these multiplets is equivalent to an understanding of the symmetries and spectrum generating algebras of the string.

In the simplest theories in twenty-six and ten dimensions, the multiplets are representations of the Heisenberg algebra or its supersymmetric extension, and the number of states at each mass level is given by the relevant partition function.

In the standard compactification to lower dimensions first introduced in 1976 by Cremmer, Scherk and Schwarz [2], there was superimposed on the Regge recurrences with mass^2 splitting given by the slope parameter, a Kaluza - Klein tower of states with mass splitting scaled by the inverse of the radius of compactification R. In a new compactification scheme (a light-cone version of the Frenkel-Kac construction) described in these notes, we set $\alpha'=R^2$, thereby increasing the degeneracy at a given level of mass. With just a few more restrictions we can force the spectrum of the compactified open string to lie in an irreducible representation of an affine Kac Moody Lie algebra. It is the basic representation and is such that the massless particles carry an adjoint representation of a finite parameter

Lie group. In the case of closed strings this is a gauge symmetry and is much larger than the isometry group of the internal space.

Section II reviews the affine algebras and their representations. In earlier work[3], it was shown that Kaluza-Klein field theory carries a representation of the loop algebra (an affine algebra with no central extension) and has the same number particles at each mass level. The representations of affine algebras (with non zero central term), which occur only in the quantum theory, have a highest weight, and the degeneracy at each mass level mushrooms. It is this mushrooming we are seeing in the spectra of compactified string theories. Even the free string spectrum is a quantum mechanical system.

A new quantum number can be identified for particles which lie in a representation of an affine algebra. We call this the affinecharge[4]. It is the eigenvalue of the derivation operator and counts the grade of representation. In the Kaluza-Klein field theory loop algebra, the affinecharge takes on all integer values and corresponds to the electric charge. In compactified dual string theories, the affinecharge is non-negative integers and corresponds to the contribution to the mass squared from the internal dimensions. That is to say there are now two quantum numbers whose sum is the total mass.

Section III describes a general compactification scheme[5] which reduces the free Veneziano open string by an arbitrary number of dimensions. For each helicity, the spectrum forms basic representations of affine G where rank G is equal to the number of internal dimensions. These results are presented in part as the first step in the construction of a four-dimensional interacting string theory. The results have also been applied to the closed heterotic string[6], and the gauge representation content of its particle spectrum is calculated to the fourth level.[5]

Section IV reviews the connection between compactification of the open string Veneziano model on the twenty-four dimensional Leech lattice and representations of the Monster group[7,8].

II. Affine Kac-Moody Lie Algebras and their Representations

In these lectures, we consider the class[9] of "untwisted" affine

Kac-Moody Lie algebras $\hat{G} = G \times C[t,t^{-1}] + cC + dC$. Here G is a finite parameter Lie algebra. $\hat{G}$ has an infinite number of generators, and irreducible representations of $\hat{G}$ are infinite dimensional consisting of an infinite tower of representations of G.

The commutation relations of $\hat{G}$ can be written as:

$$[T^n_a, T^m_b] = if_{abe} T^{n+m}_e + cn\, \delta_{n,-m}\, \delta_{ab}$$
$$[c, T^n_a] = 0$$
$$[d, T^n_a] = T^n_a$$
$$[c,d] = 0 \;. \qquad (1)$$

Here n is an integer, d is the derivation operator, $T_a{}^0$ are the generators of G, f_{abe} are the structure constants of G, c is a central operator (c commutes with all the other generators and can therefore be thought of as a c-number), and $c\, n\, \delta_{n,-m}\delta_{ab}$ is called the central extension. Eq. (1) satisfies the Jacobi identity for any value of c, but in order for there to exist highest weight representations of (1), c is restricted to particular real values.

The Cartan subalgebra of $\hat{G}$ is $\{H_I, c, d\}$ where H_I is the Cartan subalgebra of G. In the Cartan-Weyl basis, linear combinations of the elements $T_a{}^0$ are written as H_I and the step operators E_α where

$$[H_I, H_J] = 0$$
$$[H_I, E_\alpha] = \alpha^I E_\alpha \;. \qquad (2)$$

$I = 1,\ldots r$, r is the rank of G, and the ((dim G)-r) different α^I are the roots of G. These can be expressed as linear combinations with integer coefficents of the r simple roots α^I_L, $\alpha^I = \sum_{L=1}^{r} N^L \alpha_L{}^I$.

The Cartan matrix of G is $C_{LM} = 2\, \dfrac{\vec{\alpha}_L \cdot \vec{\alpha}_m}{(\vec{\alpha}_m)^2}$.

In the Cartan-Weyl basis, the elements of $\hat{G}$ can be written as $\{H_I, c, d\}$ and the step operators $H_I{}^n$ $(n \neq 0)$ and $E_\alpha{}^n$:

$$[H_I, c] = [H_I, d] = 0$$

$$[H_I, E_\alpha{}^n] = \alpha_n{}^I E_\alpha{}^n, \; [c, E_\alpha{}^n] = 0, \; [d, E_\alpha{}^n] = nE_\alpha^n$$

$$[H_I, H_J^n] = 0, \; [c, H_J{}^n] = 0, \; [d, H_J^n] = n\, H_J^n \quad . \tag{3}$$

The roots associated with $H_I{}^n$ and $E_\alpha{}^n$ are $(0,0,n)$ and $(\alpha^I,0,n)$ respectively. The r+1 simple roots are $(\alpha_L{}^I,0,0)$ and $(-\lambda^I,0,1)$where λ^I is the highest root of G. As in the finite parameter case, the inner product is derived from:

$$g_{AB} = \mathrm{Tr}\, \bar{H}_A \bar{H}_B = \begin{pmatrix} \delta_{IJ} & 0 \\ 0 & \begin{matrix} 01 \\ 10 \end{matrix} \end{pmatrix} \tag{4}$$

where $\bar{H}_A \equiv \{H_I, 2c, d\}$. Here Tr is evaluated by considering Tr cc = Tr c $[T_a^1, T_a^{-1}]$ = Tr$[c, T_a^1]\, T_a^{-1}$= 0 etc. The equation Tr dd=0 is an extra condition which can be imposed since in Eq.(1) d is defined only up to an arbitrary shift by c.

The Cartan matrix of $\hat{G}$ can now be evaluated to be:

$$C_{\bar{L}\bar{M}} = \begin{pmatrix} C_{LM} & \dfrac{-2\Sigma\alpha_L^I\, \lambda^I}{(\vec{\lambda})^2} \\ \dfrac{-2\Sigma\lambda^I\, \alpha_M^I}{(\vec{\alpha}_M)^2} & 2 \end{pmatrix} \tag{5}$$

A feature of the affine Lie algebras is that det $C_{\bar{L}\bar{M}} = 0$, in contrast to the property det $C_{LM}>0$ which is true for all finite dimensional semi-simple complex Lie algebras.

The quantum numbers of a state are the eigenvalues of the Cartan subalgebra, since this is the maximum set of commuting generators and therefore simultaneously diagnalizable operators. These eigenvalues are also called weights. In the case of $\hat{G}$, the highest weight of a given representation is labelled by $\vec{\mu} = (\mu_I, 2c, \mu_d)$ so that $\{H_I, 2c, d\}$ $|h.w.\rangle = \vec{\mu}\,|h.w.\rangle$. The highest weight state $|h.w.\rangle$ is defined such that all step operators, $H_I{}^n$, $E_\alpha{}^n$ $n \geq o$ (corresponding to the positive roots of $\hat{G}$) annihilate $|h.w.\rangle$.

The highest weight irreducible representations of $\hat{G}$ are classified by a set of r+1 non-negative integers q_L called the Dynkin labels:

$$q_L = \left(\frac{2\,\Sigma\,\alpha_L^I \mu^I}{(\vec{\alpha}_L)^2} \quad , \quad \frac{2(-\Sigma \lambda^I \mu^I + c)}{(\vec{\lambda})^2} \right) \tag{6}$$

Since each q_L is a non-negative integer, c can take on the allowed values

$$2c = 2\,\lambda^I \mu^I - 2\,\frac{\Sigma\,\alpha_L^I \mu^I}{(\vec{\alpha}_L)^2}(\vec{\lambda})^2 + \;0,1,2,\ldots \tag{7}$$

The vectors α_L^I and λ^I are fixed for a given G. Therefore, <u>for a particular allowed value of c</u> there will be only a finite number of different choices for μ_I i.e. for possible highest weight states. That is to say, for a given $\hat{G}$ with a given value of c, there is a finite number of different highest weight representations. Thus if c can be evaluated in some physical theory, only a finite number of representations of $\hat{G}$ will be available, in contrast to the situation for G.

The highest weight representations for c≠o are infinite dimensional and they have the feature that each of them is a tower of representations of G. The tower is a stack of different "grades", each grade being distinguished by its affinecharge. For example, for $\hat{G}=\hat{SU}(2)$,

$$q_1 = 2\mu_1 \tag{8a}$$

$$q_2 = -2(\mu_1 + c). \tag{8b}$$

Adding 8 (a,b), we find 2c = 0,1,2,... and that for a given c, there are 2c +1 highest weight representations. For any $\hat{G}$, the highest weight representation with the lowest non-zero value of c and with $\mu_I=0$ i.e. |h.w.> is a singlet of G, is called <u>the basic representation</u> of $\hat{G}$. For $\hat{SU}(2)$ and c=1/2, the two highest weight representations have |h.w.> equal to an SU(2) singlet ($q_1=0$, $q_2=1$) and an SU(2) doublet ($q_1=1$, $q_2=0$) respectively. The SU(2) content can be examined as follows:

We define the step operators $T_\pm^n = T_1^n \pm i\, T_2^n$. Then T_-° lowers SU(2) isospin by a one unit and leaves the affinecharge (eigenvalue of d) unchanged. T_+^{-1} raises isospin and lowers the affinecharge by one unit. The basic representation thus has $|h.w.\rangle = |\mu_1 = 0, \mu_d = 0\rangle$ in the zero grade of the tower, $T_+^{-1}|h.w.\rangle$, $T_-^\circ T_+^{-1}|h.w.\rangle$, and $T_-^\circ T_-^\circ T_+^{-1}|h.w.\rangle$ an SU(2) triplet at the first grade. Other states such as $T_+^\circ T_+^{-1}|h.w.\rangle$ vanish since they can be checked to have zero norm. This pattern is worked out for the first few grades in terms of SU(2) representations:

c = 1/2

1

3

3,1

3,3,1

5,3,3,1,1

etc. (9)

The c=1/2 highest weight doublet has

2

2

2,4

etc. (10)

For c = 1, (q_1 = 2, q_2 = 0) is a highest weight adjoint:

3

3,1

5,3,3,1

etc. (11)

For $\hat{SU}(3)$, the basic representation has

1

8

8,8,1

10,$\overline{10}$,8,8,8,1,1

27,10,$\overline{10}$,8,8,8,8,8,8,1,1,1

etc. (12)

A feature of the basic representations of $\hat{G}$ is that the first grade after the singlet is always the adjoint representation of G. The G content of a given representation can be expressed by multiplicity formulas[11]. For example, $n_k(s)$ is equal to the number of SU(2) multiplets with highest weight k in grade s for ($q_1 = 0$, $q_2 = 1$):

$$\phi_k(x) \equiv \sum_{s=0}^{\infty} n_k(s) x^s = x^{k^2}(1-x^{2k+1}) \prod_{\ell=1}^{\infty} \frac{1}{(1-x^\ell)} \qquad k=0,1,2\ldots \tag{13}$$

III. Dual String Models

Uncompactified Strings

In dual string models, elementary particles are associated with vibrational modes of a one-dimensional string in a higher-dimensional space. As the string moves in time, it traces out a two-dimensional world sheet described by $\chi^\mu(\sigma,\tau)$. The single particle states of the dual model are the quantum states of a massless relativistic string, i.e. the Fock space of D massless two dimensional fields $\chi^\mu(\sigma,\tau)$, $\mu=0,..D-1$, whose action and equations of motion[12] are

$$S = -\frac{1}{2\Pi\alpha'} \int_0^{\Pi} d\sigma \int_{\tau_i}^{\tau_f} d\tau [(\partial_\sigma \chi^\mu \partial_\tau \chi_\mu)^2 - (\partial_\sigma \chi)^2 (\partial_\tau \chi)^2]^{1/2} \tag{14}$$

$$(\partial_\sigma^2 - \partial_\tau^2)\, \chi^\mu(\sigma,\tau) = 0 \tag{15}$$

where the reparameterization invariance of (14) has been used to choose the coordinate conditions

$\partial_\sigma \chi \cdot \partial_\tau \chi = 0$, $(\partial_\sigma \chi)^2 + (\partial_\tau \chi)^2 = 0$ and $\chi^+ = x^+ + 2\alpha' p^+ \tau$ where $\chi^\pm = \frac{\chi^0 \pm \chi^{D-1}}{\sqrt{2}}$.

For open string boundry conditions $\partial_\sigma \chi^\mu(0,\tau) = \partial_\sigma \chi^\mu(\pi,\tau)$, the normal mode expansion solves (15):

$$\chi^\mu(\sigma,\tau) = x^\mu + 2\alpha' p^\mu \tau + i\sqrt{2\alpha'} \sum_{n\neq 0} \frac{1}{n} \alpha_n^\mu \cos n\sigma\, e^{-in\tau}. \tag{16}$$

Here α' is the slope parameter with dimension mass^{-2}. From (16) we see that a string $X^\mu(\sigma,\tau)$ is an infinite superposition of quantum modes $\alpha_n{}^\mu$, where a specific quantum mode corresponds to a point particle.

In a covariant quantization, the operators $\alpha_n{}^\mu$ satisfy a Heisenberg algebra:

$$[\alpha_n{}^\mu, \alpha_m{}^\nu] = n\delta_{n,-m}\eta^{\mu\nu}$$

$$p^\mu = \frac{1}{\sqrt{2\alpha'}}\alpha_o^\mu$$

$$[x^\mu, p^\nu] = i\eta^{\mu\nu}$$

$$[x^\mu, \alpha_n{}^\nu] = 0 \text{ for } n\neq 0$$

$$\eta^{\mu\nu} = \text{diag}\,(-1,1,\ldots 1)$$

$$\mu = \pm, i\ ;\quad i=1,\ldots D-2. \tag{17}$$

Additional useful combinations are defined by

$$N \equiv \sum_{n=1}^{\infty} \alpha_{-n}{}^\mu \alpha_{n\mu}$$

$$L_m \equiv \frac{1}{2}\sum_{n=-\infty}^{\infty} :\alpha^\mu_{m-n}\alpha_{n\mu}:,\quad L_o = \alpha' p^2 + N$$

$$D^{-1} \equiv p^2 + M^2 = L_o - 1$$

$$\alpha' M^2 \equiv N-1$$

$$[L_n, L_m] = (n-m)\, L_{n+m} + \frac{D}{12}\,(n^3-n)\,\delta_{n,-m}\,. \tag{18}$$

A consistent interacting quantum string theory will require D=26. The SO(25,1) Lorentz algebra is defined as

$$M^{\mu\nu} = x^\mu p^\nu - x^\nu p^\mu - i\sum_{n=1}^{\infty}\frac{1}{n}\,(\alpha^\mu_{-n}\alpha_n^\nu - \alpha^\nu_{-n}\alpha_n{}^\mu). \tag{19}$$

The Fock space states are labelled by their SO(24) content, $\alpha' M^2$, and p^μ. The highest weight state is a tachyon $|0\rangle$: $\alpha_n{}^\mu|0\rangle=0$ for $n>0$. Then $\alpha' M^2\,(|0\rangle) = -1(|0\rangle)$ and $\alpha' M^2(\alpha_{-1}{}^\mu|0\rangle)=0$, but $\alpha_{-1}{}^\mu|0\rangle$ is not on shell since $[p^\mu, \alpha_n{}^\nu]=0$. Also, since $\langle 0|\,\alpha_1{}^0\,\alpha_{-1}{}^0\,|0\rangle = -1$, $\alpha_{-1}{}^0|0\rangle$ is a ghost (a negative norm state).

Ghosts are eliminated by imposing the coordinate conditions on the physical Hilbert space. This is the requirement that $L_n|\text{phys}\rangle = 0$ for $n>0$.

To construct physical particle states (on shell states with positive norm), we define a vertex operator

$$V(k,z) = e^{\left(\sqrt{2\alpha'}\, k^\mu \sum_{n=1}^{\infty} \frac{1}{n} \alpha_{-n}{}^{\mu} z^n\right)} e^{ik.x + 2\alpha' k.p \ln z}$$

$$e^{\left(-\sqrt{2\alpha'}\, k^\mu \sum_{n=1}^{\infty} \frac{1}{n} \alpha_n^\mu z^{-n}\right)} \tag{20}$$

and $$P^\mu(z) = \frac{1}{\sqrt{2\alpha'}} \sum_{n=-\infty}^{\infty} \alpha_n{}^\mu z^{-n} \tag{21}$$

and the DDF operators [13]

$$A_n^i = \frac{1}{\sqrt{2}} \frac{1}{2\pi i} \oint \frac{dz}{z} P^i(z)\, V(n\, k_0, z) \tag{22}$$

where $k_0^\mu = \dfrac{-1}{2\sqrt{2\alpha'}p+} \quad (1,0,..0,-1).$

Then, from Eq. (17), we find

$$[p^-, A_n{}^i] = \frac{-n}{2\alpha' p+} A_n^i \tag{23a}$$

$$[\alpha' M^2, A_n^i] = -nA_n{}^i \tag{23b}$$

$$[A_n{}^i, A_m{}^j] = n\delta_{n,-m}\delta^{ij}. \tag{24}$$

Eq. (23a) should be contrasted with

$$[p^-, \alpha_n^\mu] = 0. \tag{25}$$

Therefore $A_{-1}^i|0\rangle$ is <u>on shell</u> and has positive norm for $i = 1,...24$. In general, the physical states are in a representation of the Heisenberg algebra $[A_n{}^i, A_m{}^j] = n\delta_{n,-m}\delta^{ij}$, which is affine $U(1)^{24}$, and are given by $|0\rangle$, $A_{-1}^i|0\rangle$, $A_{-1}^i A_{-1}^j|0\rangle$, $A_{-2}^i|0\rangle$ etc. The number of states $n(s)$ at level $\alpha' M^2 = s-1$ is given by the partition function:

$$q(\eta(q))^{-24} = \prod_{\ell=1}^{\infty} (1-q)^{-24} = \sum_{s=0}^{\infty} n(s)q^s \ . \tag{26}$$

In light cone gauge quantization[14], some of the operators x^μ, α_n^μ themselves are eliminated by the coordinate conditions. The independent operators are $x^\pm$, x^i, p^+, p^i, A_n^i. The dependent operators are (sum on i):

$$A_n^+ = 0 \text{ for } n \neq 0 \tag{27a}$$

$$A_n^- = \frac{1}{\sqrt{2\alpha' }p^+} \frac{1}{2} \sum_{m=-\infty}^{\infty} A_{n-m}^i A_m^i \text{ for } n \neq 0 \tag{27b}$$

$$p^- = \frac{1}{2\alpha' p^+} (\alpha'(p^i)^2 + \dot{N} - 1) \tag{27c}$$

$$N \equiv \sum_{n=1}^{\infty} A_{-n}^i A_n^i \ . \tag{27d}$$

From (27c) the mass operator is

$$\alpha' M^2 \equiv 2\alpha' p^+ p^- - \alpha(p^i)^2 = N-1.$$

The commutators of the independent variables are

$$[A_n^i, A_m^j] = n\delta_{n,-m} \delta^{ij} \tag{29a}$$

$$[p^+, x^-] = i \tag{29b}$$

$$[x^i, p^j] = i \delta^{ij} \tag{29c}$$

otherwise 0.

From (29) and (27),

$$[p^-, A_n^i] = \frac{-n}{2\alpha' p^+} A_n^i \tag{30a}$$

$$[\alpha' M^2, A_n^i] = -nA_n^i \tag{30b}$$

$$[p^-, x^i] = \frac{-i}{p^+} p^i \ . \tag{31}$$

Note that in contrast with Eq.(31), in covariant gauge $[p^-, x^i] = 0$. Eq. (30) are the same commutation relations as were derived in Eq. (23) for the DDF operators in covariant gauge, thus motivating why we chose to call the <u>independent</u> operators $A_n{}^i$ rather than $\alpha_n{}^i$ in light cone quantization. However, it should be understood that the $A_n{}^i$ of Eq. (29) and the $A_n{}^i$ of Eq. (22) are completely different objects, the former being operators in light cone gauge and the latter being functions of covariant gauge operators $\alpha_n{}^\mu$.

In light cone quantization, the physical states are given by $|0\rangle$, $A_{-1}{}^i|0\rangle$, $A_{-1}{}^i A_{-1}{}^j|0\rangle$, $A_{-2}{}^i|0\rangle$ etc. Although on shell, these states have fixed components of momentum in terms of the momentum of $|0\rangle$ due to (29) and (30). A state at a given mass level with arbitrary transverse momentum k^i is given (for example for $\alpha' M^2=0$) by

$$e^{ik^i \hat{x}^i} A_{-1}^j |0\rangle \tag{32}$$

since $[\alpha' M^2, x^i] = 0$.

<u>Compactified Strings</u>

We now compactify the open bosonic string in light cone gauge. The independent degrees of freedom are given by:

$$X^i(\sigma,\tau) = x^i + 2\alpha' p^i \tau + i\sqrt{2\alpha'} \sum_{n\neq 0} \frac{1}{n} A_n^i \, e^{-in\tau} \cos n\sigma. \tag{33}$$

The position coordinate $x^i = X^i(\pi/2, 0)$ lives in a 24-dimensional space. For future reference we write $\mu = \{\pm, (i=1,\ldots D-2)\}$, $i = \{(\hat{i}=1,\ldots D-d-2), (I=1,\ldots d)\}$. Compactification of d space dimensions is the requirement that x^I lives on a d-dimensional hypertorus. In order to describe a d torus, we introduce the concept of a lattice. A d-dimensional lattice is a set of points which are integer multiples of the basis vectors of the lattice.

$$\Lambda^I = \sum_{L=1}^{d} N^L \alpha_L^I \tag{34}$$

Here N^L are integers and the basis vectors are chosen to have length^2 $= \sum_{I=1}^{d} \alpha_L^{\ I}\ \alpha_L^{\ I} = 2$ for each L. (See Figure 1 for the 2-dimensional square lattice). An arbitrary point x^I can be expressed as $x^I = \Sigma b^L \frac{\alpha_L^{\ I}}{\sqrt{2}}$ for arbitrary coefficient b^L. The condition that x^I lives on a d-torus with all radii equal to R is that the eigenvalues

$$x^I = \sum_L b^L \frac{\alpha_L^{\ I}}{\sqrt{2}} = \sum_L (b_L + 2\pi R N^L) \frac{\alpha_L}{\sqrt{2}} = x^I + \sqrt{2}\pi R \Sigma N^L \alpha_L^I. \tag{35}$$

Since $\hat{p}^I$ generates translations $[\hat{x}^I, \hat{p}^J] = i\delta^{IJ}$,

$$|x^J\rangle = e^{ix^I \hat{p}_I}|x^I=0\rangle = |x^I + \sqrt{2}\pi R\ \ \Sigma N^L \alpha_L^{\ I}\rangle = e^{(i\sqrt{2}\pi R \Sigma N^L \alpha_L^{\ I} \hat{p}_I)}|x^I\rangle. \tag{36}$$

Therefore the allowed values of the internal momentum are given by:

$$p^I = \frac{\sqrt{2}}{R}\ \Sigma N_L\ \alpha^L{}_I \tag{37}$$

where $\Sigma \alpha_L^{\ I} \alpha^L{}_J = \delta^I{}_J$ defines the basis vectors $\alpha^L{}_I$ of the dual lattice $\Lambda_I{}^* = \sum N_L \alpha^L{}_I, N^L$ integers. The dual lattice is the set of all points whose inner product with any point on the original lattice is an integer:

$$\Lambda^I \Lambda_I{}^* = \Sigma N^L N_M \alpha_L^{\ I}\ \alpha^M{}_I = \Sigma N^L N_L\ . \tag{38}$$

A lattice is integral if the product of any two points is an integer. Since, for $C_{LM} = \sum \alpha_L{}^I \alpha_m{}^I$,

$$\Lambda^I\ \Lambda^{I'} = \Sigma N^L N'^M \alpha_L^{\ I}\ \alpha_M^{\ I} = \Sigma N^L N'^M\ C_{LM} \tag{39}$$

then Λ^I is an integral lattice whenever the basis vectors $\alpha_L{}^I$ are the simple roots of a semi-simple Lie algebra G (the elements of the Cartan matrices are integers). In this case rank G = d and $\Lambda^I = \Lambda_R^I$ is

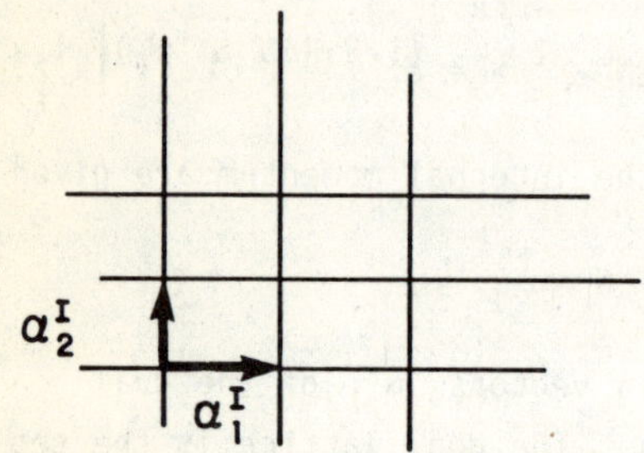

Fig. 1 - The $\mathbb{Z}^2$ lattice

called the root lattice, $\Lambda_2{}^I$ is the set of roots, $\Lambda_2{}^I \subset \Lambda_R{}^I$, and $\Lambda_{*I} = \Lambda_W$ the weight lattice (this follows since any weight μ_I of any representation of G can be written as integer coefficients times the simple roots in the dual basis

$$a_L \equiv C_{LM}\alpha^M \text{ and } \mu_I = \Sigma N^L a_L = \Sigma N^L C_{LM}\alpha^M_I = \Sigma N_M \alpha^M_I = \Lambda_{*I} \text{).}$$

If we now require x^I to live on a weight lattice, so that

$$x^I = x^I + 2\sqrt{2}\pi R\ \Sigma\ N_L\ \alpha^L_I \tag{40}$$

then

$$\sqrt{2}\ Rp^I = \Sigma\ N^L \alpha_L{}^I = \Lambda^I_R\ . \tag{41}$$

In 26-d dimensions, the Lorentz group is SO(25-d,1). The particle states are labelled by SO(24-d), p^I, $p^{\hat{i}}$, $p^{\pm}$, $\alpha' M^2_{26-d}$.

$$\alpha' M^2_{26-d} = \alpha' M^2 + \Sigma\ \alpha' (p^I)^2 = N-1 + \frac{\alpha'}{R^2}\left(\frac{1}{2} N^L N^M C_{LM}\right)\ . \tag{42}$$

For G equal to the simply laced (roots of equal length) Lie algebras $A_n = SU(n+1)$, $D_n = SO(2n)$, $E_n = E_6, E_7, E_8$, the Cartan matrix is symmetric so $N^L N^M C_{LM}$ always an even integer. The allowed values of $\sqrt{2}\ Rp^I$ given by the points on $\Lambda_R{}^I$ can be grouped into Weyl orbits, each of which is a set weights related by Weyl reflections which leave the root lattice invariant. All weights in a Weyl orbit have the same length^2 = $N^L N^M C_{LM}$. For example, for SU(3); $\alpha_1{}^I = (\sqrt{2},0)$, $\alpha_2{}^I = \left(\frac{-1}{\sqrt{2}}, \frac{\sqrt{3}}{\sqrt{2}}\right)$. (See Figures 2 and 3). The first few orbits are:

$$
\begin{array}{llll}
[0,0] = 0 & \equiv \Delta & \text{length}^2 = 0 & \\
[1,1] = \sqrt{2}Rp_1{}^I & \equiv X & \text{length}^2 = 2 & \\
[3,0] = \sqrt{2}Rp_2{}^I & \equiv 0 & \text{length}^2 = 6 & \\
[0,3] = \sqrt{2}Rp_2{}^I & \equiv \bullet & \text{length}^2 = 6 & (43)
\end{array}
$$

The orbit decomposition for the SU(3) representations is

$$
\begin{aligned}
1 &= [0,0] \\
8 &= [1,1] + 2\,[0,0] \\
10 &= [3,0] + [1,1] + [0,0] \\
\overline{10} &= [0,3] + [1,1] + [0,0] \\
&\text{etc.}
\end{aligned}
\tag{44}
$$

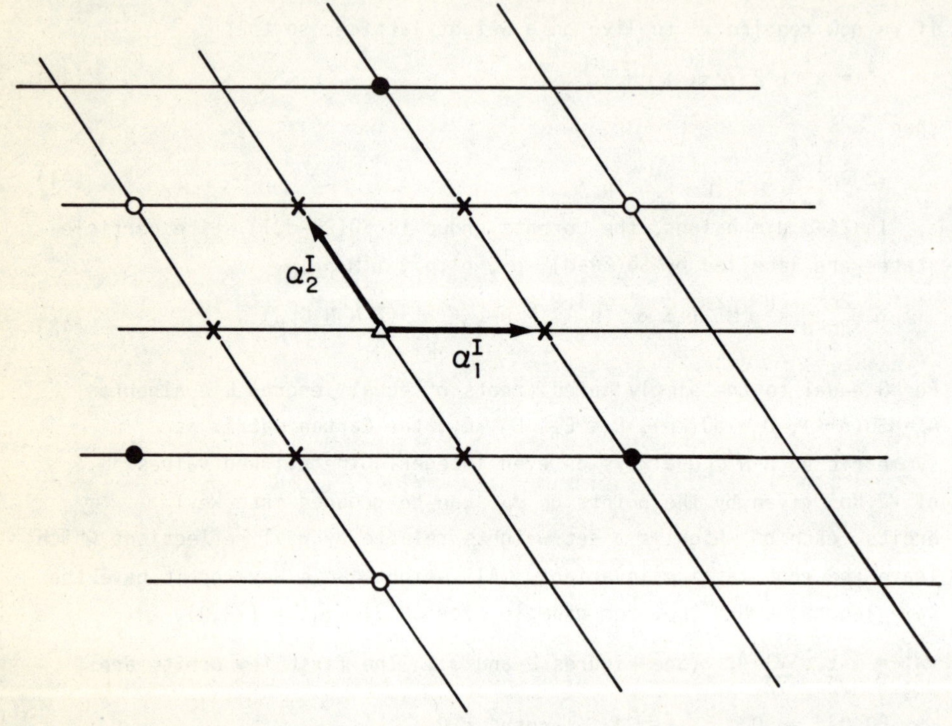

Fig. 2 - Λ_R for SU(3)

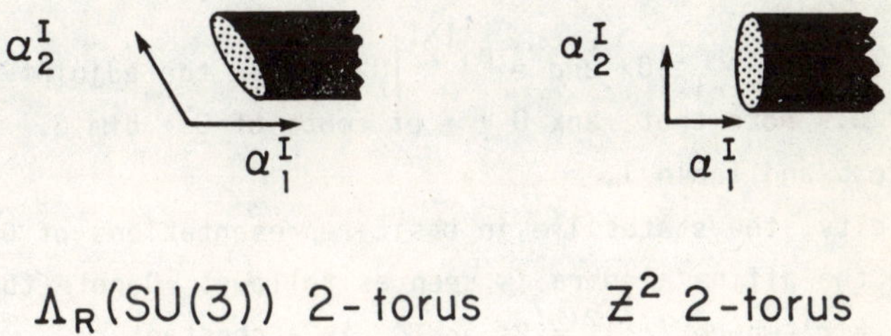

Fig. 3 - Root lattice vs. cubic lattice compactification

In the compactified string model, such states as $e^{ip^I\hat{x}^I}|0>$ have $\alpha' m^2_{26-d} = -1+\alpha'(p^I)^2$ different from the $\alpha' M^2_{26-d}$ of $|0>$, so they must be included in the spectrum. The physical states are thus given by the Regge states $|0>$, d scalars $A_{-1}^I\,|0>$, one SO (24-d) vector $A_{-1}^{\hat{i}}|0>$ etc. as well as the Kaluza-Klein states $e^{ip^I\hat{x}^I}|0>$, $e^{ip^I\hat{x}^I}\, A_{-1}^I|0>$ etc. where p^I is restricted to lie on some lattice designated by the choice of the internal space torus. See Figure 4. For $\alpha' \neq R^2$, and $\alpha_L{}^I = \sqrt{2}\ \delta_L{}^I$ then the d massless scalars $A_{-1}^I|0>$ are in $U(1)^d$. For $\alpha' = R^2$, and $\alpha_L I$ = simple roots of G = A_n, D_n, E_n then the Kaluza-Klein tower moves onto the Regge tower and the massless scalars $A_{-1}^I|0>$ and $e^{ip_1^I\hat{x}^I}|0>$ are in the adjoint representation of G. Note that rank G + # of roots of G = dim G. See for example Figure 5 and Table 1.

At fixed helicity, the states lie in basic representations of $\hat{G}$. The connection to the affine algebra is seen as follows. Denote the roots by $\sqrt{2\alpha'}\ p_1{}^I = r^I$ where $(r^I)^2 = 2$, and C_r is a constant.[5]

$$C_r\, e^{ip_1^I\hat{x}^I}|0> \;=\; X_{-1}(r)\,|0> \tag{45}$$

$$X_n(r) = \frac{C_r}{2\pi i}\oint \frac{dz}{z}\ z^n\ V(r,z)$$

$$V(r,z) = e^{r^I\sum_{n=1}^{\infty}\frac{A_{-n}^I z^n}{n}}\ e^{\frac{ir^I\hat{x}^I}{\sqrt{2\alpha'}}}\ e^{\sqrt{2\alpha'}r^I\hat{p}^I\ln z}\ z\ e^{-r^I\sum_{n=1}^{\infty}\frac{A_n^I z^{-n}}{n}}. \tag{46}$$

Here A_n^I, $\hat{x}^I$, $\hat{p}^I$ are light-cone operators[5] and $X_{-1}(r)\,|0>$ is on shell.[5]

[Note that had we kept the covariant operators α_n^I in (46), then

$$[\mathcal{L}_n, V[\alpha]] = z^n(z\frac{d}{dz} + n\frac{r^2}{2})\ V[\alpha]\ , \tag{47}$$

so that $X_{-1}(r)\,|0>$ <u>in the covariant formulation is not an on shell physical state</u>.]

In light-cone gauge, $[p^-, X_n(r)] = \frac{-n}{2\alpha' p^+}\, X_n(r)$.

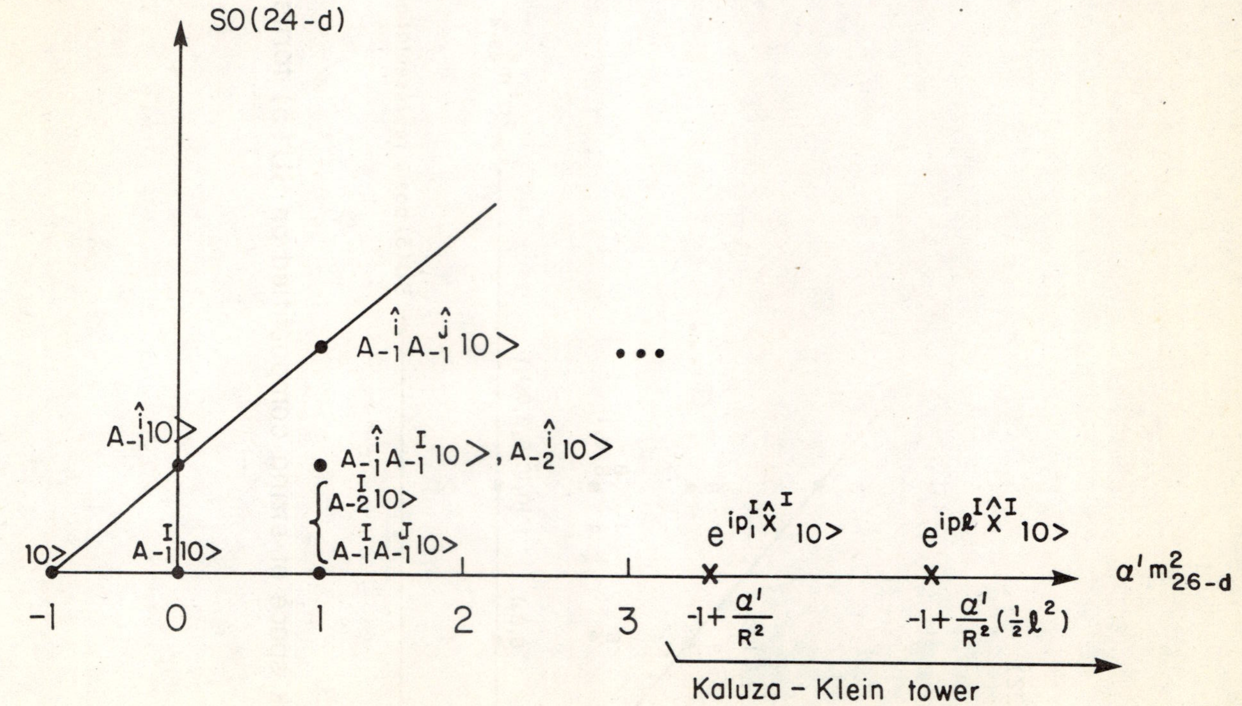

Fig. 4 - Fock space of compactified string

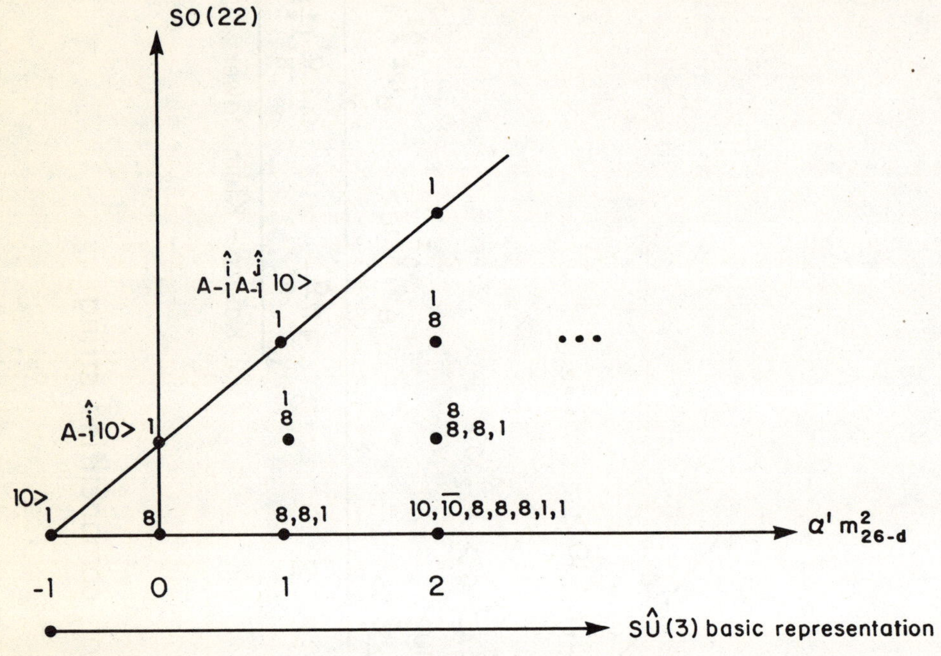

Fig. 5 - Fock space of string compactified on SU(3) torus

Table 1 - States for Figure 5

$\alpha' M^2_{22}$	State	SO(22)	SU(3)	SO(23)	SU(3)
-1	$\|0\rangle$	1	1		
0	$A^I_{-1}\|0\rangle$ $e^{ip^I_1 \hat{x}^I}\|0\rangle$	1	2[0,0]+[1,1] = 8		
0	$A^{\hat{i}}_{-1}\|0\rangle$	22	1		
1	$A^I_{-1}A^J_{-1}\|0\rangle$ $A^I_{-2}\|0\rangle$ $A^I_{-1}e^{ip^I \hat{x}^I}\|0\rangle$	1	5[0,0]+2[1,1] = 8+8+1	1	1+8
1	$A^{\hat{i}}_{-1}A^I_{-1}\|0\rangle$ $A^{\hat{i}}_{-1}e^{ip^I_1 \hat{x}^I}\|0\rangle$	22	8	23	8
1	$A^{\hat{i}}_{-1}A^{\hat{j}}_{-1}\|0\rangle$	1+252	1	275	1
	$A^{\hat{i}}_{-2}\|0\rangle$	22	1		

The operators $A_n(r) = \Sigma r^I A_n^{\ I}$ and $X_n(r)$ close to form $\hat{G}$:

$$
\begin{aligned}
[A_n(r), A_m(s)] &= 2cnr.s\ \delta_{n,-m} \\
[A_n(r), X_m(s)] &= r.s\ X_{n+m}\ (r) \\
[X_n(r), X_m(-r)] &= A_{n+m}(r)\ +2c\ n\ \delta_{n,-m} \\
[X_n(r), X_m(s)] &= \begin{cases} \varepsilon(r,s)\ X_{n+m}(r+s) & \text{if } r+s \text{ is a root} \neq 0. \\ 0 & \text{if } r+s \text{ is not a root.} \end{cases}
\end{aligned} \tag{48}
$$

Eqs.(46) and (48) are thus a light-cone version[5] of the Frenkel-Kac construction[15] of the basic representation of $\hat{G}$ for simply laced G.

Furthermore, we can check that the gauge algebra $A_0{}^I$ and $X_0(r)$ commutes with the Lorentz algebra in the uncompactified dimensions, SO(25-d,1). This is essential in order that the SO(24-d) multiplets at a given massive level which are now in various gauge representations can be combined into irreducible SO(25-d) multiplets while respecting the gauge symmetry.

The above construction can be applied to the left-movers of the closed heterotic string[6]. The allowed values of momentum are $\sqrt{2\alpha'}\ p^I = \Lambda^I$ is either the root lattice of $E_8 \times E_8$, or the Spin $32/Z_2$ weight lattice whose points are Λ_R of SO(32) and $\Lambda_S = \Lambda_R + \lambda_S$, $\lambda_S = (1/2, 1/2, \ldots 1/2)$. The physical states of the $E_8 \times E_8$ heterotic string are given by direct products of right-moving, N=1 superstring states and the left-moving bosonic states consisting of the standard SO(8) multiplets of the Heisenberg U(1)[8] algebra, each of which is the highest weight of the c=1/2 representation of affine $E_8 \times E_8$ given in Table 2. The squared masses are four times those given in the table, and the tachyon is decoupled due to the closed string constraint condition.

The physical states of the Spin $32/Z_2$ heterotic string are given in the same manner except that now there are two c=1/2 representations of affine SO(32), one highest weight singlet and one highest weight spinor. Since $\alpha'(p^I)^2 = 1/2\ \ell^2{}_{sp} = 2$ for the length$^2 = \ell^2{}_{sp}$ of the spinor weight, the highest weight spinor representation first appears at $\alpha' M^2{}_{10} = 1$.

TABLE 2: The first four levels of the $E_8 \times E_8$ compactitifed open bosonic string.

$\alpha' M^2_{10}$	No. States	$E_8 \times E_8$ representation labelled by dimensionality.
-1	1	(1,1)
0	496	(248,1) + (1,248)
1	69752	(3875,1) + (1,3875) + (248,248) + (248,1) + (1,248) + 2(1,1)
2	2115008	(30380,1) +(1,30380) +(3875,248) + (248,3875) + 2(248,248) + (3875,1) + (1,3875) + 3(248,1) + 3(1,248) + 2(1,1)
3	34670620	(147250,1) + (1,147250) + (27000,1) + (1,27000) + (30380,248) + (248,30380) + (3875,3875) + (30380,1) + (1,30380) + 2(3875,248) + 2(248,3875) + 3(3875,1) + 3(1,3875) + 5(248,248) + 5(248,1) + 5(1,248) + 5(1,1)

IV. Leech Lattice, Monster Group, and the Affine Construction

In carrying out the affine construction we saw that it was useful to know how many different values of the allowed internal momenta have the same length^2 $(2\alpha'((p^I)^2)$. This information is contained in a modular form, the theta function[16] of the lattice $\theta_\Lambda(q)$, where the expansion $\theta_\Lambda(q)=\sum_{q=0}^{\infty} c_n q^n$ gives c_n as the number of points on the lattice Λ with $(\Lambda^I)^2 = 2\alpha'(p^I)^2 = 2n$. The E_8 root lattice has[16]:

$$\theta_{E_8}(q) = 1+240 \sum_{n=1}^{\infty} \sigma_3(n)q^n$$

$$= 1+240q + 2160q^2 + 6720q^3+\ldots \tag{49}$$

Here $\sigma_3(n)$ is the sum of cubes of positive divisors of n. In the construction of the basic representation of E_8, the number of states n(s) at each grade is thus given by the character of that representation which called V_Z:

$$\mathrm{ch}V_Z = \frac{\theta_{E_8}(q)}{(\eta(q))^8} = q^{-\frac{1}{3}} \sum_{s=0}^{\infty} n(s)q^s.$$

$$= q^{\frac{-1}{3}} \quad (1+248q + 4124q^2+\ldots) \tag{50}$$

Here $(\eta(q))^{-d}$ is the partition function for the Heisenberg algebra states in d-dimensions. See Eq.(26). The first few grades of the $\hat{E}_8$ c=1/2 representation have E_8 content 1; 248; 1+248+3875; etc. which is in agreement with Eq. (50).

It is crucial in the construction of the affine generators and the derivation of their commutation relations Eq.s (48) that $\sum (r^I)^2 = 2$ and that $\sum r^I r'^I$ be an integer. Furthermore, in order that the Fock space of compactified string states (including those with internal momenta given by all points on a lattice $\Lambda^I = \sqrt{2\alpha'}p^I$) is equivalent to the states of the basic representation generated by the affine algebra, it is sufficient to require Λ^I to be an integral lattice.

However, the Fock space of compactified string states can still be constructed when the internal momentum is restricted to an arbitrary lattice. Therefore, we can consider the 26-dimensional Veneziano open string compactified on the 24-dimensional Leech lattice[17].

If we now concentrate on the internal sector of the Fock space only, i.e., on those states generated by $\Pi A^I_{n_\ell} e^{ip^J_j \hat{x}^J} |0\rangle$, for $\sqrt{2\alpha'}\, p^I_j = \Lambda^I_{\text{Leech}}$, $I = 1,\ldots 24$, then we are eventually led to a (reducible) representation of the largest sporadic group, the Monster F_1. The theta function for the Leech lattice is given by[16]

$$\theta_L(q) = 1 + \frac{65520}{691}\left[\left(\sum_{n=1}^{\infty} \sigma_{11}(n) q^n\right) - (\eta(q))^{24}\right]$$

$$= 1 + 196{,}560 q^2 + 16{,}773{,}120 q^3 + \ldots \tag{51}$$

The number of states n(s) at each mass level is given by

$$\frac{\theta_L(q)}{(\eta(q))^{24}} = q^{-1} \sum_{s=0}^{\infty} n(s) q^s$$

$$= q^{-1}\,(1 + 24q + 196{,}884 q^2 + 21{,}493{,}760 q^3 + \ldots). \tag{52}$$

Apart from the 24q term, Eq.(52) is the normalized modular function:

$$J(q) = q^{-1} + 196{,}884 q + 21{,}493{,}760 q^2 + \ldots \tag{53}$$

The coefficients in this expansion are linear combinations of the dimensions of the irreducible representations of the Monster: 196,883; 21,296,876; etc., a fact included in the set of ideas known as "monstrous moonshine" and the "monstrous game".

In order to actually construct representations of the Monster in the string Fock space language, the states described by Eq.(52), namely those outlined in Table 3,I., are modified by combining some of them with the states of Table 3, II. of a <u>twisted</u> construction[7]. Starting

Table 3 - Leech lattice compactified string states - natural module for F_1

$\alpha' M^2$	-1	0	$\frac{1}{2}$	1	$\frac{3}{2}$	2	...
I.	$\|0>$	$\cancel{A^I_{-1}\|0>}$		$A^I_{-1}A^J_{-1}\|0>(300)$		$\cancel{A^I_{-1}A^J_{-1}A^K_{-1}}\|0>(2600)$	
				$\cancel{A^I_{-2}\|0>}$ (24)		$\cancel{A^I_{-3}\|0>}$ (24)	
				$\|p_2^+>$ (98280)		$A^I_{-1}A^J_{-2}\|0>$ (576)	
				$\cancel{\|p_2^->}$ (98280)		$\cancel{A^I_{-1}\|p_2^+>}$ (24x98280)	
						$A^I_{-1}\|p_2^->$ (24x98280)	
						$\|p_3 + >$ (8,386,560)	
						$\cancel{\|p_3 - >}$ (8,386,560)	
II.			$\cancel{\|2^{12}>}$	$c^I_{-\frac{1}{2}}\|2^{12}>(98304)$	$\cancel{c^I_{-\frac{1}{2}}c^J_{-\frac{1}{2}}\|2^{12}>}$	$c^I_{-\frac{1}{2}}c^J_{-\frac{1}{2}}c^K_{-\frac{1}{2}}\|2^{12}>$	
						2600.2^{12}	
						$c^I_{-\frac{3}{2}}\|2^{12}>=(24\text{x}2^{12})$	

$$|p^{\pm}> \equiv \pm |(-p)^{\pm}>$$

from an arbitrary even lattice, one can introduce twisted vertex affine Lie algebras. For an automorphism of order 2, the Heisenberg operators $A_n{}^I$, $n \epsilon Z$ are replaced by half-integrally moded operators [18] $c_s{}^I$, $s \epsilon Z + 1/2$:

$$[c_s^I, c_u^J] = s \delta_{s,-u} \delta^{IJ} \qquad (54)$$

$$V(r,z) = e^{r^I \sum_{s=1/2}^{\infty} \frac{c_{-s}^I z^s}{s}} \quad e^{-r^I \sum_{s=1/2}^{\infty} \frac{c_s^I z^{-s}}{s}} \qquad (55)$$

$$r^I \ \epsilon \ \text{roots of a rank } d \text{ simply laced } G, \ \sum_I (r^I)^2 = 2. \qquad (56)$$

The twisted affine algebra $\hat{G}_{\text{twisted}}$ has generators $X_n(r)$, n = integer or half-integer, and c_s^I:

$$X_n(r) = \frac{1}{2\pi i} \gamma \oint \frac{dz}{z} z^{2n} V(r,z^2).$$

In this case $X_n(r) = X_n(-r)$ for $n \epsilon Z$ and $X_n(r) = -X_n(-r)$ for $n \epsilon Z + 1/2$. γ plays a role similar to the cocycle of Eq.(45). The generators $X_0(r)$ for $r>0$ close a subalgebra G_0 of G. A property of the irreducible representations of $\hat{G}_{\text{twisted}}$ is that they are towers not of G representations but of G_0 representations, and thus may provide a mechanism for symmetry breaking in the string model.

Since there is no zero-mode momentum operator in the set $c_s{}^I$, the Fock space is given by a "highest weight" spinor state $|\alpha\rangle$ whose dimension is $2^{d/2}$, and $c_{-1/2}^I |\alpha\rangle$, $c_{-1/2}^I c_{-1/2}^J |\alpha\rangle$, etc.

For this twisting and $G=E_8$, then $G_0 = SO(16)$ and the Fock space is $|16\rangle$, $c_{-1/2}|16\rangle = |128\rangle, \ldots$ which forms two irreducible $c=1/2$ representations of $\hat{SO}(16)$: the highest weight vector $|16\rangle$ and a highest weight spinor $|2'\rangle$.

For d=24, the states are outlined in Table 3.II. The construction of the "natural module" corresponds to keeping states from I. and II. which are even under $p \to -p$, $A \to -A$ and odd under $c \to -c$.

The physical relevance of the connection between natural modules of compactified strings and representations of the sporadic groups is currently under examination.

The most useful property of the representation theory reviewed in these notes is that a light-cone version of the affine vertex operator construction leads to a non-abelian gauge invariant, Lorentz invariant compactified dual string model. This is due to the general feature of the basic representation of $\hat{G}$ that the first grade below the highest is the adjoint of G, and that the group generators are constructed as the contour integral of a vertex operator whose conformal spin is 1, thereby insuring that the generators of G commute with the Lorentz algebra of the remaining umcompactified space-time. Extensions of this algebraic construction to other string models is under current investigation.

In the spirit of unified string theory and its uniqueness, namely the extremely restrictive mathematical origins of the models, it is possible that the further connection with sporadic groups may play a fundamental physical role.

References

1. For recent reviews see J. Schwarz, Cal Tech preprint 88-1247 and references therein; and M.B. Green, Surveys in H.E. Physics 3, 127 (1983).
2. J. Scherk and J. Schwarz, Phys. Lett. B57, 463 (1975). E. Cremmer and J. Scherk, Nuc. Phys. B103, 399 (1976).
3. L. Dolan and M.J. Duff, Phys. Rev. Lett. 52, 14 (1984); and L.Dolan, in Ref. 7, p. 353.
4. L. Dolan, APS-DPF Meeting at Sante Fe, November, 1984, unpublished.
5. L. Dolan and R. Slanksy, Phys. Rev. Lett. 54, 2075 (1985).
6. D.J. Gross, J.A. Harvey, E. Martinec and R. Rohm, Phys. Rev. Lett 54, 502 (1985).
7. For an introduction see for example I. Frenkel, J. Lepowsky, and A. Meurman in Vertex Operators in Mathematics and Physics, ed. by J. Lepowsky, S. Mandelstam and I. Singer p. 231 (Springer-Verlag, New York, 1985).
8. B. Griess, p. 217, ibid.
9. For other physics-style reviews of affine algebras (which also list the original mathematical references) see L. Dolan, Phys.Rep. 109, (1984); P. Goddard and D. Olive, Lectures delivered at the Srni Winter School, January 1985.
10. J. Lepowsky and G.Wilson, Com. Math. Phys. 62, 43 (1978).
11. V.G. Kac and D.H. Peterson, Adv. in Math. 53, 125 (1984).
12. For reviews see S. Mandelstam, Phys.Rep.13C, 259 (1974) J. Schwarz, Phys. Rep. 8C, 269 (173) and 89C,223 (1982).
13. E.DelGuidice, P. DiVecchia, and S. Fubini, Ann. Phys. (NY) 70, 378 (1972).
14. J. Goldstone, P. Goddard, C. Rebbi, and C. Thorn, Nuc. Phys. B56, 109 (1973).
15. I. Frenkel and V.G. Kac, Inv. Math. 62, 23(1980); G. Segal,Com. Math. Phys. 80, 301 (1981); I. Frenkel, Amer. Math. Soc. Lectures in Appl. Math. 21, 235 (1985); P. Goddard and D. Olive, in Ref. 7, p. 51.

16. J.P. Serre, A Course in Arithmetic (Springer-Verlag, New York 1973).

17. This discussion follows the mathematical discussion of Ref. 7 and references therein. It is included here to show the similarity with compactified string theory.

18. E. Corrigan and D. Fairlie, Nucl.Phys. B91, 527(1975).

TORSION on the WORLD-SHEET

Thomas Curtright

Physics Department, University of Florida, Gainesville, FL 32611

Seminar given at the Lewes Workshop on Superstrings, 12 July 1985.

ABSTRACT

We discuss the dynamical consequences of introducing torsion into nonlinear σ models defined on two-dimensional spacetime. We then describe corresponding properties of the covariant world-sheet formulation for the freely propagating heterotic superstring. For chiral theories, including covariant superstrings, we summarize some of the arguments which show how a particular strength for the torsion yields, on two-dimensional world sheets, an ultraviolet finite and trivially integrable theory (i.e. free oscillators).

1. CHIRAL NONLINEAR σ MODELS

In general, nonlinear σ models are maps, $\phi(x)$, from spacetime into a field manifold, M, which is usually taken to be Riemannian. The dynamics of the field is specified by an action which depends on the geometry of M. The conventional "metric term" for the action is

$$I_g = (2\lambda^2)^{-1} \int d^2x \, g_{ab}(\phi) \, \delta^{\mu\nu} \, \partial_\mu\phi^a \, \partial_\nu\phi^b \tag{1.1}$$

where g_{ab} is the metric of M, and where λ^2 may be regarded as a loop counting parameter. The resulting equations of motion for the field define a natural connection on the manifold, namely that of Levi-Civita: $\Gamma_{abc} = (\partial_b g_{ac} + \partial_c g_{ab} - \partial_a g_{bc})/2$. Using $\Gamma^a{}_{bc} \equiv g^{ad}\Gamma_{dbc}$, the equations of motion are simply an expression of covariant constancy of a contravariant vector. Thus

$$0 = D_\mu \partial^\mu \phi^a \equiv (\delta^a{}_b \, \partial_\mu + \Gamma^a{}_{cb} \, \partial_\mu \phi^c) \, \partial^\mu \phi^b \tag{1.2}$$

where D_μ is the usual covariant derivative on M, pulled back to spacetime.

As may easily be seen, all isometries of the metric are symmetries of the dynamics. That is, if $\mathcal{L}_\xi g_{ab} \equiv - D_a\xi_b - D_b\xi_a = 0$, then $\delta\phi^a = \xi^a$ is an invariance of the action (1.1).

In two-dimensional spacetime, there is another possible contribution to the action which is quadratic in spacetime derivatives of the field [1—4]. It requires an antisymmetric tensor on the manifold, e_{ab}. The appropriate form is

$$I_e = \eta\, (3\lambda^2)^{-1} \int d^2x\, e_{ab}(\phi)\, \epsilon^{\mu\nu}\, \partial_\mu\phi^a\, \partial_\nu\phi^b \tag{1.3}$$

and in general, it has arbitrary strength, η, relative to the metric term in (1.1). I_e is the "torsion term" in the action, since its presence modifies the natural definition of the connection for the model by inducing a totally antisymmetric torsion component [5], the curl of e_{ab}.

$$S_{abc} \equiv \frac{1}{3}\eta\, (\partial_a e_{bc} + \partial_b e_{ca} + \partial_c e_{ab}) \tag{1.4}$$

Hence, e_{ab} is the "torsion potential". Again, the natural definition of the connection is revealed through the equations of motion, obtained by varying $I_g + I_e$. The result is similar to (1.2).

$$0 = \mathcal{D}_\mu\partial^\mu\phi^a \equiv D_\mu\partial^\mu\phi^a + S^a{}_{cb}\, \epsilon^{\mu\nu}\, \partial_\nu\phi^c\, \partial_\mu\phi^b \tag{1.5}$$

and involves the pull-back of a covariant derivative with torsion.

Previous isometries of the metric are not necessarily symmetries of I_e, although they were of I_g. In order to be a symmetry of both contributions to the action, an infinitesimal isometry must induce a "gauge transformation" on e_{ab}. That is, $\delta\phi^a = \xi^a$ is a symmetry of I_e only if $\mathcal{L}_\xi e_{ab} = \partial_a\zeta_b - \partial_b\zeta_a$ where $\mathcal{L}_\xi e_{ab} \equiv -(\partial_a\xi^c)e_{cb} - (\partial_b\xi^c)e_{ac} - \xi^c\partial_c e_{ab}$.

All of the above concepts are explicitly realized by chiral σ models, for which the field manifold is the coset space G×G/G, where G is a semisimple Lie group. For such models, the manifold may be parameterized by

$$U[\phi] \equiv \exp[i\phi^a T_a] \qquad \in \qquad G\times G/G \tag{1.6}$$

where T_a is a generator in the defining representation of G, chosen so that $[T_a,T_b] = 2\, i\, f_{abc}\, T_c$, and $\mathrm{Tr}(T_a T_b) = 2\, \delta_{ab}$. In terms of these group elements, the metric and torsion potential are given by [3,6]

$$g_{ab} = (f_{ab} + f_{ba})/2 \qquad e_{ab} = -3\; (f_{ab} - f_{ba})/4$$
$$f_{ab} \equiv \int_0^1 dt\; (1-t)\; \mathrm{Tr}\left\{ T_a\, U^t\, T_b\, U^{-t} \right\} \tag{1.7}$$

Geometrical structures on the coset space are also expressible in terms of Lie algebra valued one-forms [6], or "left-invariant vielbeine".

$$V_a{}^j\, T_j = i\, U^{-1}\, \partial_a U \qquad V_a{}^j = \frac{i}{2}\, \mathrm{Tr}\left\{ T_j\, U^{-1}\, \partial_a U \right\} \tag{1.8}$$

These obey the well-known equations of Maurer and Cartan.

$$D_a\, V_b{}^j = -\, f_{jkl}\, V_a{}^k\, V_b{}^l \tag{1.9}$$

In terms of the vielbeine, other fundamental tensors on the manifold are

$$g_{ab} = \delta_{jk}\, V_a{}^j\, V_b{}^k \qquad S_{abc} = \eta\, f_{jkl}\, V_a{}^j\, V_b{}^k\, V_c{}^l$$
$$R_{abcd} = f_{jkn}\, f_{lmn}\, V_a{}^j\, V_b{}^k\, V_c{}^l\, V_d{}^m \tag{1.10}$$

Here R_{abcd} is the conventionally defined Riemann curvature. It is also useful to define the generalized curvature including torsion.

$$\mathcal{R}_{abcd} \equiv R_{abcd} + D_c\, S_{abd} - D_d\, S_{abc} + S_{fac}\, S^f{}_{db} - S_{fad}\, S^f{}_{cb} \tag{1.11}$$

It follows from (1.9), (1.10), and the Jacobi identities for f_{jkl} that

$$D_a\, S_{bcd} = 0 \qquad \mathcal{R}_{abcd} = (1-\eta^2)\, R_{abcd} \tag{1.12}$$

Thus the torsion is covariantly constant for *any* value of η, as is g_{ab}, and the generalized curvature vanishes when $\eta = \pm 1$. The latter feature is known as "parallelism" induced by torsion.

Another way to understand parallelism is to consider the spinor connection of Cartan.

$$\omega_a{}^{jk} \equiv V^{bj}\, D_a V_b{}^k = (\eta-1)\, f_{jkl}\, V_a{}^l \tag{1.13}$$

where in the second step we used the Maurer-Cartan equations, e.g. to write

$$S_{abc} = -\,\eta\, V_a{}^j\, D_{[b} V_{c]}{}^j \tag{1.14}$$

In terms of ω, the generalized curvature is given by

$$R_{abcd} = 2\, V_a{}^j V_b{}^k \left\{ \partial_{[c}\omega_{d]}{}^{jk} + \omega_{[c}{}^{jl}\omega_{d]}{}^{lk} \right\} \tag{1.15}$$

and obviously vanishes when $\eta = 1$ since ω does. Similarly, a right-invariant vielbein may be defined which makes it obvious that $R = 0$ when $\eta = -1$.

Upon parallelism, the chiral σ model is actually a free field theory in disguise. Classically, there is a straightforward argument to see this based on geometrical considerations [6]. Consider the "vector potential" which appears in D_μ in (1.5).

$$A_\mu{}^a{}_b \equiv \Gamma^a{}_{bc}\, \partial_\mu \phi^c - S^a{}_{bc}\, \epsilon_{\mu\nu}\, \partial^\nu \phi^c \tag{1.16}$$

A direct calculation of the field strength for this potential gives

$$\begin{aligned} &\partial_\mu A_\nu{}^a{}_b - \partial_\nu A_\mu{}^a{}_b + A_\mu{}^a{}_c A_\nu{}^c{}_b - A_\nu{}^a{}_c A_\mu{}^c{}_b \\ &\quad = \epsilon_{\mu\nu} \left\{ - S^a{}_{bc}\, (D_\lambda \partial^\lambda \phi)^c + D_c S^a{}_{bd}\, (\partial_\lambda \phi^c\, \partial^\lambda \phi^d) \right\} \\ &\qquad - 3\, \epsilon_{\mu\nu}\, S^a{}_{e[b}\, S^e{}_{cd]}\, (\partial_\lambda \phi^c\, \epsilon^{\lambda\rho}\, \partial_\rho \phi^d) \\ &\qquad + \left\{ R^a{}_{bcd} - D_c S^a{}_{bd} + D_d S^a{}_{bc} \right\} \partial_\mu \phi^c \partial_\nu \phi^d \end{aligned} \tag{1.17}$$

It follows from (1.5), (1.12), and the Jacobi identities that this field strength vanishes ***on-shell*** when the manifold is parallelized. In that case, we have a pure gauge form for A_μ.

$$A_\mu = U^{-1}(x)\, \partial_\mu\, U(x) \tag{1.18}$$

where U is a ***local*** function of spacetime. Given this local U, the scalar field equations (1.5) are solved by transporting a free field.

That is, let

$$\partial_\mu \phi(x) \equiv U^{-1}(x)\, \partial_\mu \phi_0(x) \tag{1.19}$$

Then the ϕ field equations become

$$D^\mu \partial_\mu \phi(x) = (\partial^\mu + A^\mu)\, U^{-1}\, \partial_\mu \phi_0 = U^{-1}\, \partial^\mu \partial_\mu \phi_0(x) \tag{1.20}$$

which vanishes if and only if ϕ_0 is a free, massless scalar field. Since the relation in (1.19) is invertible, we conclude that upon parallelism of the field manifold, there is a local, on-shell equivalence between ϕ and a free field ϕ_0. It is reasonable to expect this equivalence to persist at the quantum level, and indeed the S-matrix has been argued to be trivial for this model [3], when $\eta = \pm 1$. The quantum theory will be discussed further in Section 3 below.

2. SUPERSYMMETRIC GENERALIZATIONS

Supersymmetric generalizations of two-dimensional σ models are easily obtained [5–9]. Majorana spinor fields ψ^a are added to the model to form supermultiplets with the scalar fields ϕ^a. The appropriate generalization of the dynamics is obtained essentially by replacing the scalar fields with real, unconstrained superfields Φ^a, which contain ϕ^a, ψ^a, and real auxiliaries F^a as the coefficients of the usual polynomial in the Majorana Grassman variable θ: $\Phi^a \equiv \phi^a + \bar{\theta}\psi^a + \bar{\theta}\theta\, F^a/2$. In addition, the supercovariant derivative of the superfield is required: $D\Phi^a \equiv \psi^a + (\, F^a - i\gamma^\mu \partial_\mu \phi^a\,)\, \theta + \bar{\theta}\theta\, i\gamma^\mu \partial_\mu \psi^a/2$.

The following manifestly supersymmetric actions then contain the general bosonic actions, with arbitrary metric g and torsion potential e.

$$\begin{aligned} I_g &= (2\lambda^2)^{-1} \int d^2x\, \frac{1}{2} \int d^2\theta\, g_{ab}(\Phi)\, \overline{D\Phi}^a\, D\Phi^b \\ I_e &= -\,\eta\, (3\lambda^2)^{-1} \int d^2x\, \frac{1}{2} \int d^2\theta\, e_{ab}(\Phi)\, \overline{D\Phi}^a\, \gamma_P\, D\Phi^b \end{aligned} \tag{2.1}$$

where the pseudoscalar matrix is $\gamma_P = \gamma^0\gamma^1$. Expanding the arbitrary functions g_{ab} and e_{ab} as second order polynomials in θ, and performing the integration over the θ variable in the usual way, the action is obtained explicitly in terms of component fields. As usual, the auxiliary fields only appear quadratically, without spacetime derivatives, and can be replaced through their equations of motion which imply the auxiliary field is nothing but the full connection contracted with a spinor bilinear. Upon eliminating F_a, the action involves only ϕ^a and ψ^a. It is then straightforward to combine the various terms into D-covariant derivatives and generalized curvatures, as defined earlier, to obtain

$$
\begin{aligned}
I_S = (2\lambda^2)^{-1} \int d^2x \Big\{ & g_{ab}\, \partial_\mu \phi^a \partial^\mu \phi^b + i\, g_{ab}\, \overline{\psi}^a\, (\gamma^\mu D_\mu \psi)^b \\
& + \tfrac{2}{3}\eta\, e_{ab}\, \epsilon^{\mu\nu}\, \partial_\mu \phi^a \partial_\nu \phi^b - \tfrac{1}{3}\eta\, \partial_\mu (i\, e_{ab} \overline{\psi}^a \gamma_P \gamma^\mu \psi^b) \\
& + \tfrac{1}{8} R_{abcd}\, \overline{\psi}^a (1+\gamma_P) \psi^c\, \overline{\psi}^b (1+\gamma_P) \psi^d \Big\}
\end{aligned} \tag{2.2}
$$

The D-covariant derivative of ψ^a is defined exactly like that of $\partial_\mu \phi^a$.

$$
(D_\mu \psi)^a \equiv \Big\{ \delta^a{}_b\, \partial_\mu + \Gamma^a{}_{cb}\, \partial_\mu \phi^c + S^a{}_{cb}\, \epsilon_{\mu\nu}\, \partial^\nu \phi^c \Big\} \psi^b \tag{2.3}
$$

Note that the spinor dependent analogue of the torsion term in (2.2) is a total divergence, and may be discarded assuming that the product of e_{ab} with the spinor fields vanishes as $|x| \to \infty$. Also note that the tensor $e_{ab}(\phi)$ in (2.2) must be the same function of ϕ as in the previous bosonic model, if the supersymmetric model is to yield that previous bosonic case upon setting $\psi = 0$. Finally, note that the quartic spinor coupling is proportional to R_{abcd} and therefore vanishes upon parallelism of the scalar field manifold, as in (1.12) when $\eta = \pm 1$.

Upon parallelism, there is an argument for the general supersymmetric σ model, completely parallel to that for the bosonic σ model, which shows the theory to be equivalent to a free field system [6,9]. In the supersymmetric case, ***both*** ϕ and ψ can be written as D-transported free fields.

To see the details of the argument again from a geometric point of view, we consider the classical equations of motion of the theory. The local variation of the general action (2.2) with respect to ψ^a gives the spinor equations of motion. Multiplying by λ^2, these are

$$\Psi^a \equiv i(\gamma^\mu D_\mu \psi)^a + \tfrac{1}{8}\left\{R^a{}_{bcd} + R_{cb}{}^a{}_d\right\}\psi^c\,\bar\psi^b(1+\gamma_P)\psi^d$$
$$+ \tfrac{1}{8}\left\{R^a{}_{bcd} - R_{cb}{}^a{}_d\right\}\gamma_P\,\psi^c\,\bar\psi^b(1+\gamma_P)\psi^d \qquad (2.4)$$

which vanish when the spinors are on-shell. Similarly, $(-\lambda^2)$ times a variation with respect to ϕ^a gives

$$\Upsilon^a \equiv (D^\mu\partial_\mu\phi)^a - i\,\bar\psi^b\psi^c\,\Gamma_{cb}{}^a - i\,\bar\psi^b\gamma_P(\gamma^\mu D_\mu\psi)^c\,S_{cb}{}^a$$
$$- \tfrac{1}{16}\,(D^a R_{bcde})\,\bar\psi^b\,(1+\gamma_P)\,\psi^d\,\bar\psi^c\,(1+\gamma_P)\,\psi^e$$
$$- \tfrac{1}{2}\,i\,\bar\psi^b\,(\gamma^\mu\partial_\mu\phi^d)\,\psi^c\left\{R_{bc}{}^a{}_d - D^a S_{bcd} + D_d S_{bc}{}^a\right\} \qquad (2.5)$$
$$- \tfrac{1}{2}\,i\,\bar\psi^b\,\gamma_P\,(\gamma^\mu\partial_\mu\phi^d)\,\psi^c\left\{D^a S_{bcd} - D_d S^a{}_{bc}\right\}$$

Again, this vanishes when the scalars are on-shell.

The equations of motion reduce considerably when $R_{abcd} = 0 = D_a S_{bcd}$, i.e. in the case of parallelism. In that case, the on-shell fields satisfy

$$\gamma^\mu D_\mu\psi = 0 = D^\mu\partial_\mu\phi \qquad \textit{(upon parallelism)} \qquad (2.6)$$

Thus, if the potential A_μ is defined in terms of Γ and S exactly as in (1.16), its field strength is given exactly as in (1.17), and hence vanishes given parallelism (with $0 = R_{abcd} = D_a S_{bcd} = S_{a[bc}S_{d]ef}$) and the equations of motion in (2.6). The vanishing of the field strength for A_μ then allows it to be expressed as in (1.18) in terms of a ***local*** exponential $U(x)$. Using this exponential, we then define local fields ϕ_0 and ψ_0 by

$$\partial_\mu\phi_0(x) \equiv U(x)\,\partial_\mu\phi(x) \qquad \psi_0(x) \equiv U(x)\,\psi(x) \qquad (2.7)$$

It then follows from that ϕ_0 and ψ_0 are free fields. This establishes the local equivalence of the classical supersymmetric σ model, upon parallelism, to a free field system. As in the case of the purely bosonic model, it is not unreasonable to expect that this equivalence obtains also in the quantum theory, to be briefly discussed next.

3. QUANTUM EFFECTS

Interesting dynamical effects arise from the torsion term in the action. Perhaps the most dramatic are quantum renormalization effects. Under a change of length scale, the geometry of the model changes — the metric and torsion potentials evolve [10—12,5,6]. To two-loop order, this evolution is given by the following renormalization group equations.

$$M \frac{d}{dM} \left\{ \frac{1}{\lambda^2} g_{ab} + \frac{2\eta}{3\lambda^2} e_{ab} \right\} = \frac{1}{2\pi} R_{ab} + \frac{\lambda^2}{6\pi^2} R_{zxya} R_b{}^{(xy)z} \qquad (3.1)$$

This result is valid for the purely bosonic model. For the supersymmetric model, the $O(R^2)$ terms are absent [12]. It is not known whether the RHS of (3.1) is a polynomial in the generalized curvature R when extended to higher orders, although it seems very likely that this is the case. It is also not known whether there are higher order corrections for the supersymmetric case, but again it seems plausible that there are no corrections, at least to three loops.

It is known [13] that upon parallelism of the manifold, which essentially restricts the theory to be a chiral σ model, the extension of the RHS of (3.1) vanishes to all orders of perturbation theory. Thus the geometry of the model ceases to change with length scale upon parallelism, an effect known as "geometrostasis" [5,6]. For the chiral models defined earlier, along with their supersymmetric generalizations, it seems that parallelism also implies that the full quantum theory reduces to free fields, as suggested by the classical field analysis given above, although this has not been rigorously established in all cases [14—16]. Finally, it is possible [13] to argue that the extension of the antisymmetric part of the RHS of (3.1) vanishes to all orders for chiral σ models, even when the manifold is not parallelized (i.e. even when η is not ± 1). This is explicitly true to two-loop order, as may be checked using (3.1) and the properties of chiral models given above. Thus the torsion contribution I_e is evidently not renormalized for general chiral models. This result is consistent with topological arguments which require the torsion term to be quantized [1—4,6].

4. HETEROTIC SUPERSTRING

The heterotic superstring [17] may be viewed as a hybrid between right- and left-moving sectors of the bosonic string in 26 dimensions and the N=1 superstring [18] in ten dimensions. The world-sheet action may be written covariantly using projective coordinates for the coset space (N = 1 supersymmetry in ten dimensions) mod (Lorentz group in ten dimensions), along with 16 = 26−10 world-sheet bosonic variables arising from the mismatch in critical dimensions for the two sectors. Alternatively, an equivalent way to incorporate the extra 16 variables is to use 32 "ancillary" Majorana-Weyl world-sheet spinors.

In terms of the latter variables, the covariant world-sheet action for the heterotic superstring is a sum of three terms: $I_{HS} = I_g + I_e + I_a$.

$$I_g = (2\pi)^{-1} \int d^2x\, (-g)^{1/2}\, g^{\alpha\beta}\, \delta_{\mu\nu}\, E_\alpha{}^\mu E_\beta{}^\nu$$

$$I_e = (2\pi)^{-1}\, 2i \int d^2x\, \epsilon^{\alpha\beta}\, \partial_\alpha X^\mu\, (\bar\theta\, \Gamma_\mu\, \partial_\beta \theta) \tag{4.1}$$

$$I_a = (2\pi)^{-1}\, i \int d^2x\, (-g)^{1/2}\, P_+{}^{\alpha\beta}\, \bar\psi^j\, \gamma_\alpha\, \partial_\beta \psi^j$$

Here X^μ is a ten-dimensional spacetime coordinate, but a world-sheet scalar; θ is also a world-sheet scalar, but a Majorana-Weyl spinor in ten dimensions; Γ_μ are Dirac matrices in ten dimensions; $g^{\alpha\beta}$ is a metric for the world-sheet which is in general a curved two-dimensional surface; ψ^j (j=1,⋅⋅,32) is a scalar in ten dimensions, but a world-sheet Majorana spinor; and finally, $P_\pm{}^{\alpha\beta} \equiv [\, g^{\alpha\beta} \pm (-g)^{-1/2}\, \epsilon^{\alpha\beta}]/2$ are projection operators, one of which enforces the two-dimensional Weyl condition on ψ^j. That is, $P_+{}^{\alpha\beta}\, \gamma_\alpha\, \partial_\beta \psi^j \equiv \gamma^\alpha \partial_\alpha\, (1+\gamma_P)\, \psi^j/2$.

Also appearing in (4.1) are the super-one-form components, $E_\alpha{}^\mu$, which are obtained from $U^{-1}dU$, where $U \equiv \exp[\, iP_\mu X^\mu + \bar Q_m \theta^m]$ is an element of the super coset space, (N=1 supersymmetry in ten dimensions) mod SO(9,1). Spinor-valued one-form components, $E_\alpha{}^m$, are obtained in this manner as well.

$$E^M \equiv (E^\mu , E^m) = (dX^\mu - i\, \bar{\theta}\, \Gamma^\mu\, d\theta \, , d\theta^m)$$
$$E^M = E_\alpha{}^M\, dx^\alpha \tag{4.2}$$

In fact, the structure of $I_g + I_e$ is exactly what one would expect for a σ model defined on the super coset space [19], with both metric (I_g) and torsion (I_e) interaction terms. Indeed, the relative strength of these two terms is precisely that which gives vanishing generalized curvature in the super coset space, and is therefore analogous to the geometrostatic limit of the previously discussed σ models [20]. This latter point follows from the superspace analogue of the Maurer-Cartan equations in (1.9).

$$dE^\mu = - i\, (C\Gamma^\mu)_{mn}\, E^m\, E^n \tag{4.3}$$

where C is the charge conjugation matrix in ten dimensions. The coefficients appearing on the RHS of (4.3) are of course just the structure constants of the N = 1 supersymmetry algebra.

As in the previous σ model cases, the equations of motion provide a natural definition of the connection on the superspace. For example, in the conformal gauge, where $(-g)^{1/2}\, g^{\alpha\beta} = \delta^{\alpha\beta}$, the equations of motion for E^μ are

$$0 = \partial^\alpha E_\alpha{}^\mu + i\, \bar{E}_\alpha\, \Gamma^\mu\, E_\beta\, \epsilon^{\alpha\beta} \tag{4.4}$$

The analogy with the earlier σ models defined on parallelized manifolds goes even further, in that the dynamics defined by (4.1) yields a set of free oscillators on the world-sheet. This is a consequence of a local (on the world-sheet) ten-dimensional fermionic symmetry [18] whose parameters are $(\kappa_\alpha)^m$, i.e. world-sheet vectors, but ten-dimensional Majorana-Weyl spinors. The infinitesimal transformations are

$$\begin{aligned} \delta\theta &= i\, \Gamma_\mu\, E_\alpha{}^\mu\, P_-{}^{\alpha\beta}\, \kappa_\beta & \qquad \delta X^\mu &= i\, \bar{\theta}\, \Gamma^\mu\, \delta\theta \\ \delta e^{\alpha a} &= - P_-{}^{\alpha\beta}\, \partial_\beta \bar{\theta}\, e_\gamma{}^a\, P_-{}^{\gamma\delta}\, \kappa_\delta & \qquad \delta\psi^j &= 0 \end{aligned} \tag{4.5}$$

Here $e_\alpha{}^a$ is the zweibein for the world-sheet, with the world-sheet metric given by $g_{\alpha\beta} = e_\alpha{}^a\, e_\beta{}^a$ Note that $I_g + I_e$ and I_a are separately invariant under (4.5). Also note that the ancillary world-sheet spinors ψ^j are inert under this local transformation, just as they were inert under the global super coset space translations in ten spacetime dimensions. Exploiting all the local symmetries of the theory, and solving all constraints, the theory can be formulated in the light-cone gauge with: $X^0+X^9=2^{1/2}\,P^+\tau$, $(\Gamma^0+\Gamma^9)\theta=0$, $(-g)^{1/2}g^{\alpha\beta}=\delta^{\alpha\beta}$, and $(e_\tau{}^0-e_\sigma{}^0)+(e_\sigma{}^1-e_\tau{}^1)=2$. The resulting equations of motion describe free oscillators, and can be derived from the light-cone action for the heterotic string.

$$
\begin{aligned}
I_{HS}(\textit{light-cone}) &= (2\pi)^{-1}\int d^2x\,\tfrac{i}{2}\,\overline{\psi}^j(\gamma^0+\gamma^1)(\partial_\sigma+\partial_\tau)\psi^j \\
&\quad + (2\pi)^{-1}\int d^2x\left\{-\,\partial_\alpha X^t\partial^\alpha X^t+\tfrac{i}{2}\,\overline{\theta}\,(\Gamma^0-\Gamma^9)(\partial_\sigma-\partial_\tau)\theta\right\}
\end{aligned}
\tag{4.6}
$$

which involves the transverse coordinates, X^t, $t=1,\cdot\cdot,8$. The extension of this discussion to the case of the superstring in curved ten-dimensional N = 1 superspace is given in [21], while the corresponding discussion of the N = 2 case is given in [22].

ACKNOWLEDGEMENTS

My understanding of and contributions to this subject developed in collaborations with Cosmas Zachos, Eric Braaten, and Luca Mezincescu. I thank them for sharing their expertise. This work was supported by the U.S. Department of Energy (Contract No. DE-AS-05-81ER40008).

REFERENCES

[1] J. Wess and B. Zumino, Phys. Lett. **37B** (1971) 95.

[2] S.P. Novikov, Sov. Math. Dokl. **24** (1981) 222.

[3] A. Polyakov and P. Wiegmann, Phys. Lett. **131B** (1983) 121; Phys. Lett. **141B** (1984) 223.

[4] E. Witten, Commun. Math. Phys. **92** (1984) 455.

[5] T.L. Curtright and C.K. Zachos, Phys. Rev. Lett. **53** (1984) 1799.

[6] E. Braaten, T.L. Curtright, and C.K. Zachos, Nucl. Phys. B (in press).

[7] J. Gates, C. Hull, and M. Rocek, Nucl. Phys. **B248** (1984) 157.

[8] P. Howe and G. Sierra, Phys. Lett. **148B** (1984) 451.

[9] R. Rohm, Princeton preprint, October 1984.

[10] D. Friedan, Phys. Rev. Lett. **45** (1980) 1057.

[11] L. Alvarez-Gaume, D.Z. Freedman, and S. Mukhi, Ann. Phys. (NY) **134** (1981) 85.

[12] B.E. Fridling and A.E.M. van de Ven, Stony Brook preprint, ITP-SB-85-30, 1985.

[13] S. Mukhi, Tata Institute preprints, TIFR-TH-85-12 and 13, 1985.

[14] P. Goddard, W. Nahm, and D. Olive, Phys. Lett. **160B** (1985) 111.

[15] S.P. de Alwis, Bristol preprint, 1985.

[16] I. Antoniadis, C. Bachas, C. Kounnas, and P. Windey, preprint SLAC-PUB-3789, 1985.

[17] D. Gross, J. Harvey, E. Martinec, and R. Rohm, Phys. Rev. Lett. **54** (1985) 340; Nucl. Phys. **B256** (1985) 253.

[18] M. Green and J. Schwarz, Phys. Lett. **136B** (1984) 367; Nucl. Phys. **B243** (1984) 285.

[19] M. Henneaux and L. Mezincescu, Phys. Lett. **152B** (1985) 340.

[20] T.L. Curtright, L. Mezinescu, and C.K. Zachos, Phys. Lett. B (in press).

[21] E. Witten, Princeton preprint, 85-0458, 1985.

[22] M.T. Grisaru, P. Howe, L. Mezincescu, B. Nilsson, and P.K. Townsend, Texas preprint, UTTG-15-85, 1985.

WHEN IS THE CENTRAL CHARGE OF THE VIRASORO ALGEBRA IN STRING THEORIES IN CURVED SPACETIME NOT A NUMERICAL CONSTANT ?

A.N.Redlich

Department of Physics
Brandeis University
Waltham, MA 02254

Abstract

The path integral approach to string quantization is used to discuss the conditions under which the central charge c of the Virasoro algebra for string theories in curved space-time is not a numerical constant. If the internal space for the non-linear σ model -- from which the string is constructed -- is a general Riemannian manifold with torsion, and if the σ model contains only kinetic and torsion terms, then if c is to be a numerical constant, the scalar curvature R(X) (not including torsion) must be a constant function of its coordinates X.

Introduction

I am going to begin this talk with a brief review of the path-integral approach to string quantization, as it has been developed by Polyakov[1]. The standard action for a free bosonic string,

$$I = -\frac{T}{2}\int d^2\sigma \det{}^{1/2}(\partial_a X_\mu \partial_b X^\mu) \tag{1}$$

is proportional to the area of the world sheet swept out by the string as it moves through space-time. $X^\mu(\sigma^1,\sigma^2)$ represent the coordinates of d-dimensional Minkowski space, $\mu=0,\ldots,d-1$; they can be thought of as fields defined on the two-dimensional parameter space (σ^1,σ^2) of the world sheet. An alternative action which is classically equivalent to (1)) is

$$I = -\frac{T}{2}\int d^2\sigma \sqrt{h}\, h^{ab}(\sigma)\, \partial_a X_\mu \partial_b X^\mu \tag{2}$$

with $h_{ab}(\sigma)$ a two-dimensional metric on the world sheet.

This action is manifestly invariant under reparametrization of the world sheet $\sigma^{i'} = \sigma^{i'}(\sigma^{i})$. The gravitational field $h_{ab}(\sigma)$ is classically not a dynamical field; its equations of motion serve only to impose the constraints

$$T_{ab} = \partial_a X^{\mu} \partial_b X_{\mu} - \tfrac{1}{2} h_{ab} h^{cd} \partial_c X^{\mu} \partial_d X_{\mu} \tag{3}$$

or

$$h_{ab} = \partial_a X^{\mu} \partial_b X_{\mu} \tag{4}$$

Inserting (4) into (2) gives back the original action (1).

By choosing the gauge $h_{ab}=e^{\phi}\eta_{ab}$ -- η_{ab} is the two-dimensional Minkowski metric -- one discovers that the gravitational field decouples from the action (2), since the combination $\sqrt{h}\, h_{ab}=\eta_{ab}$: classically the conformal mode ϕ decouples. Quantum mechanically, however, the conformal mode does not necessarily decouple because there exists a conformal (trace) anomaly

$$\langle T^{\mu}{}_{\mu} \rangle = \frac{d}{24\pi} R_h(\sigma) \quad , d = \text{space-time dimension} \tag{5}$$

where $R_h(\sigma)$ is the two-dimensional scalar curvature coming from h_{ab}. Classically, of course, the theory is conformally invariant and $T^{\mu}{}_{\mu}=0$. To take the conformal mode ϕ into account Polyakov defines the string partition function to include a path integral over the gravitational field h_{ab} :

$$Z = \int [dh_{ab}(\sigma)] \; e^{-\lambda \int d^2\sigma \sqrt{h}} \; e^{-F[h]} \tag{6}$$

$$\exp - F[h] = \int [dX^{\mu}(\sigma)] \cdot \exp -\frac{1}{2} \int d^2\sigma \sqrt{h}\, h^{ab} \partial_a X^{\mu} \partial_b X_{\mu}$$

The effective action $F[h]$ can be calculated completely by using the conformal anomaly (5):

$$\frac{2}{\sqrt{h}} h^{ab} \frac{\delta F}{\delta h^{ab}} = h^{ab} \langle T_{ab} \rangle = \frac{d}{24\pi} \left[R_h(\sigma) + \text{constant} \right] \tag{7}$$

In the conformal gauge $h_{ab} = e^{\phi} \eta_{ab}$, we get

$$F = -\frac{d}{48\pi} \int d^2\sigma \left[\frac{1}{2} (\partial_a \phi)^2 + \mu^2 e^{\phi} \right] \tag{8}$$

with μ^2 a renormalization of the cosmological constant. In fixing the conformal gauge one must include a Faddeev-Popov determinant, which can be evaluated, giving for the measure of the gravitational field

$$[dh_{ab}(\sigma)] = [d\phi(\sigma)] \exp + \frac{13}{24\pi} \int d^2\sigma \left[\frac{1}{2} (\partial_a \phi)^2 + \mu^2 e^{\phi} \right] \tag{9}$$

Combining (8) and (9), the partition function becomes

$$Z = \int [d\phi(\sigma)] \exp \left\{ - \frac{(26-d)}{48\pi} \int d^2\sigma \left[\frac{1}{2} (\partial_a \phi)^2 + \mu^2 e^{\phi} \right] \right. \tag{10}$$

which tells us that for $d < 26$ we must include a Liouville action for the conformal mode, which greatly complicates the string theory -- for $d > 26$ the functional integral is not defined. From (10) we discover one of the most important differences between the classical and quantum theories : the conformal mode, ϕ, does not decouple quantum mechanically except under very special circumstances -- for the bosonic string when $d=26$. The conformal anomaly is therefore responsible for one of the most significant features of quantum string theories: they uniquely determine the dimension of space-time.

A similar analysis for the supersymmetric string, starting with the supersymmetric version of the action (2), gives the partition function

$$Z = \int [d\phi][d\chi] \exp\left\{ -\frac{(10-d)}{8\pi}\left[\int d^2z \; \frac{1}{2}(\partial_a\phi)^2 + \frac{1}{2} i \bar{\chi}(\gamma\cdot\partial)\chi + \frac{1}{2}\mu \bar{\chi}\gamma_5\chi e^{2\phi} + \frac{1}{2}\mu^2 e^{2\phi} \right] \right\} \quad (11)$$

(χ is what is left of the gravitino field after fixing the superconformal gauge $\chi_\alpha = \gamma_\alpha \chi$) so that the superconformal mode ϕ, χ decouples only in $d=10$ dimensions.

Generalization to Curved Space-time

Up to now, the two-dimensional fields $X^\mu(\sigma)$ have been chosen to be coordinates of d-dimensional Minkowski space, with metric $\eta^{\mu\nu}$. In general, however, they are coordinates of some manifold M with metric $g_{\mu\nu}(X^\mu(\sigma))$. The action for a free bosonic string moving in curved space-time is the obvious generalization of (2)

$$I_B = \frac{1}{2\lambda^2}\int d^2\sigma \sqrt{h}\, h^{ab}\, g^{\mu\nu}(X^\mu(\sigma))\, \partial_a X_\mu \partial_b X_\nu \qquad (12)$$

which is a non-linear σ model coupled to two-dimensional gravity. Here the coupling constant λ is related to the string tension T and the radius of the compact dimensions R by $\lambda^2 = R^2 T$.

In addition to the metric $g_{\mu\nu}(X)$ one may define a torsion potential $e_{\mu\nu}(X)$ on M and generalize[2] the string model (12) by including in the action the term

$$I_{WZ} = \frac{1}{2\lambda^2}\int d^2\sigma\, \epsilon^{ab}\, e_{\mu\nu}(X)\, \partial_a X^\mu \partial_b X^\nu \qquad (13)$$

This expression is generally covariant on the two-dimensional world sheet even though it does not contain any explicit dependence on $h_{ab}(\sigma)$: it is a world scalar without h_{ab}.

Now, given the string model, (12) plus (13), we may attempt to quantize the theory using the path integral approach outlined above for the flat space-time model. Using the action (12),(13), we may define the patition function as in (6) and attempt to calculate the effective action F[h] obtained by integrating out the space-time fields $X^{\mu}(\sigma)$. However, this is not possible for most manifolds M: Although the classical non-linear σ - models considered here, even before they are coupled to two-dimensional gravity, are conformally invariant for all manifolds M, the quantum versions of these theories are not conformally invariant for most manifolds. This happens because in renormalizing these theories the β functions for $g_{\mu\nu}(X)$ and $e_{\mu\nu}(X)$ do not vanish for general M, meaning that a dimensionful renormalization scale must be introduced which destroys the conformal symmetry of the classical theory. Since the conformal transformations are a subset of reparametrizations of the world sheet, this means that these string theories cannot be independent of how one parametrizes the world sheet. At a mathematical level, one discovers that unless the β functions. β_g and β_e vanish, the conformal modes will never decouple (for any critical dimension) because the effective action $F[h]$ will contain infinities in addition to those which give rise to the anomaly. These additional infinities cannot be cancelled by the Faddeev-Popov term coming from the measure.

The β functions for a non-linear σ model defined on a manifold M have been shown[2,3,4] to vanish for the bosonic model with torsion (12),(13) only when the <u>generalized</u> curvature

$$\mathcal{R}_{abcd} = R_{abcd} + \nabla_c S_{abd} - \nabla_d S_{abc} + S_{fac} S^f{}_{db} - S_{fad} S^f{}_{cb} \tag{14}$$

vanishes; the torsion S_{abc} is defined in terms of the torsion potential e_{ab} as $S_{abc}=ne_{[ab;c]}$. For the same model without torsion ($e_{ab}=0$), the condition for vanishing β function is $R_{abcd}=0$, R_{abcd} is the curvature coming from $g_{ab}(X)$ alone. For the supersymmetric extension of the model with torsion, the condition is $\mathcal{R}_{ij}=0$, with $\mathcal{R}_{ij}$ the generalized Ricci tensor, while for the supersymmetric model without torsion ($e_{ab}=0$), the condition is $R_{ij}=0$. These conditions are summarized in table I.

	bosonic model	supersymmetric model
with torsion $e^{\mu\nu}\neq 0$	$\mathcal{R}_{abcd}=0$	$\mathcal{R}_{ij}=0$
no torsion $e^{\mu\nu}=0$	$R_{abcd}=0$	$R_{ij}=0$

Table I

Conditions for β function(s) to vanish.

Once the conditions in table I are satisfied, the only contribution to the trace of the energy-momentum tensor $\langle T^{\mu}{}_{\mu}\rangle$ will be due to the trace anomaly --as in flat space-- and one can proceed to calculate the effective action F by integrating

$$\frac{2}{\sqrt{h}}\, h^{ab}\frac{\delta F}{\delta h^{ab}} = h^{ab}\langle T_{ab}\rangle = \frac{c(M)}{24\pi}\left[R_h(\varepsilon) + \text{constant}\right] \tag{15}$$

The only difference between this equation for general manifolds and equation (7) for flat space is that for general M the coefficient of the trace anomaly, c(M), is not necessarily equal to d, the dimension of the manifold.

If M is a direct product manifold $M=M_d\times M_c$ where M_d is d-dimensional Minkowski space, and M_c is a compact "internal" space, then using $c(M)=c(M_d) + c(M_c) = d + c(M_c)$, one learns that the critical dimension d_c for string theories on such a manifold is given by

$$d_c + c(M_c) = 26 \qquad \text{Bosonic Model} \tag{16}$$

$$d_c + \frac{2}{3}c(M_c) = 10 \qquad \text{Supersymmetric Model}$$

For the special case of $M_c = G$ =group manifold, c(G) has been calculated exactly[5,6,7,8], with the result

$$c(G) = \frac{K\,d_G}{K + c_v} \tag{17}$$

where K is an integer proportional to $R\sqrt{T}$, with R the size of the compact dimensions and T the string tension; d_G is the dimension of G, equal to the number of generators; and c_v is the second Casimir invariant of the group. The reason it is possible to calculate c(G) exactly is that one may use the current algebra for the generators of the group, together with the fact that the energy-momentum tensor T_{ab} may be expressed as a bilinear function of the currents. For general M, however, c(M) cannot be calculated exactly, and one cannot know for certain whether (16) is satisfied.

<u>When is c(M) not a Numerical Constant ?</u>

While c(M) cannot in general be calculated exactly, one may still be able to eliminate many candidate manifolds using only an approximate calculation of c(M). In addition to being the coefficient of the trace anomaly, c(M) is the central charge in the Virasoro algebra

$$-i\,[T_{\pm}(x), T_{\pm}(y)] = \frac{c(M)}{24\pi}\,\delta'''(x-y) + \cdots$$

$$[L_n, L_m] = (n-m)L_{n+m} + \frac{c}{12}(n^3-n)\,\delta_{n,-m} \qquad (18)$$

in which $T_{\pm} = \frac{1}{2}(T_{00} \pm T_{10})$ (T_{ab}=energy-momentum tensor) and the L_n are moments of T_{ab}. c(M) is not in general a numerical constant -- proportional to the identity operator -- but is a quantum mechanical <u>operator</u>. Therefore, although one may not be able to calculate c(M) exactly to show that it does or doesn't satisfy (16), one may still eliminate those manifolds for which an approximate calculation of c(M) shows it is not a numerical constant. Of course if c(M) is not even a constant then it cannot possibly satisfy (16).

If one does not keep in mind the possibility that for a given manifold, M, c(M) may fail to be a numerical constant, then one may be led to draw erroneous conclusions. For example, in the supersymmetric with no torsion (e=0) one may use general arguments to conclude

$$c(M) = \frac{3}{2}d + f(\nabla, R)$$

where d is the dimension of M and f(∇,R) is a function of

the curvature R_{abcd} and its covariant derivatives. Table I tells us that in this case we are interested only in manifolds which are Ricci flat : $R_{ij}=0$. However, it is known that for Ricci flat manifolds there exist no constant functions of R and its covariant derivatives. Therefore, if we assume c(M) is a numerical constant then we deduce that $c(M)=\frac{3}{2}d$. Unfortunately, we cannot assume c(M) is constant so it _is_ possible that $f(\nabla,R)$ is nonvanishing. No argument short of an exact calculation can show rigorously whether or not $c=\frac{3}{2}d$.

The Calculation (Bosonic model with torsion)

Let me now briefly sketch how one goes about performing an approximate calculation of c(M) for a general M (for more details see reference 9). One calculates the trace anomaly in a loop expansion, that is, in an expansion in $\hbar$, which is equivalent to an expansion in the parameter λ^2 in (12) and (13). In terms of the string theory, this is also equivalent to an expansion in $1/R\sqrt{T}$,R=radius of compact dimensions, T=string tension, which for group manifolds is equivalent to an expansion in $1/K$, see equation (17).

To perform the loop expansion, I use the background field method using normal Riemannian coordinates[3,10]. Denoting the coordinates of the compact manifold $X^i(\sigma)$ --these are two-dimensional fields -- one expands $X^i(\sigma)=X^i_0 + \pi^i(\xi^i)$ around the constant background fields X^i_0. In turn, $\pi^i(\xi^i)$ are expanded in the fields $\xi^i(\sigma)$ such that the coefficients of the expansion are generally covariant functions of X^i_0. To insure that the coefficients are generally covariant tensors, one defines ξ^i to be the tangent vector to the geodesic connecting X^i_0 and $X^i_0+\pi^i(\xi^i)$. Explicitly, one solves for $\lambda^i(t)$

$$\ddot{\lambda}^i + \Gamma^i{}_{jk}\dot{\lambda}^j\dot{\lambda}^k = 0$$

with $\lambda^i(0)=X^i_0$ and $\lambda^i(1)=X^i_0+\pi^i$ and defines $\xi^i = \dot{\lambda}^i(0)$. Since ξ^i is a tangent vector to the geodesic, it is a contravariant vector. Therefore, the coefficients of any expansion in powers of ξ^i will automatically be covariant tensors.

Expanding the bosonic action (12) in powers of

$$I_B = \frac{1}{2\lambda^2}\int d^2\sigma \Big\{ \partial_\mu\xi^i\partial^\mu\xi^j \Big(g_{ij} - \frac{1}{3}R_{iajb}\xi^a\xi^b - \frac{1}{3!}\nabla_a R_{ibjc}\xi^a\xi^b\xi^c + \frac{1}{5!}\big(-6\nabla_a\nabla_b R_{icjd} + \frac{16}{3}R_{ajb}{}^m R_{cidm}\big)\xi^a\xi^b\xi^c\xi^d \Big) \Big\} \tag{19}$$

with a similar expansion for the torsion term (13). The kinetic term in (19) $g_{ij}(X_0)=\partial_\mu\xi^i\partial^\mu\xi^j$ is diagonalized by introducing the constant Veilbeins $V^a_i(X_0)$ and making the rotation $\xi^{a'} = V^a_i(X_0)\xi^i$. I have kept terms of sufficient order in ξ^i here to calculate c(M) to three loops.

Without actually calculating the relevant graphs, one may establish for the bosonic model with torsion :

<u>Bosonic model with torsion: general R</u>

$$c(1\text{ loop}) = d$$

$$c(2\text{ loops}) = a\,R + b\,S_{abc}S^{abc} \tag{20}$$

$$c(3\text{ loops}) = a\,R_{ij}R^{ij} + b\,R_{abcd}R^{abcd} + c\,(\nabla^2 S_{abc})S^{abc} + d\,\nabla^2 R + e\,\nabla_i\nabla_j R^{ij} + f\,(S_{abc}S^{ade}S^b{}_{df}S^c{}_{ef}) + F(\nabla_i\nabla_j e_{ab})$$

$F(\nabla_i\nabla_j e_{ab})$ includes many terms which are all functions of

$\nabla_i \nabla_j e_{ab;d}$ in c(1 loop) is the dimension of M; S_{abc} is the torsion $S_{abc}=n\, e_{[ab;c]}$; and the small latin coefficients are constants, independent of the properties of the manifold M.

There is no need to calculate the two-loop graphs in order to determine the values of the coefficients a and b in c(2 loops). Instead, one uses the fact from Table I that the generalized curvature $\mathcal{R}_{abcd}$ must vanish. Since the symmetric portion of the generalized Ricci tensor is

$$\mathcal{R}_{(ij)} = R_{ij} - S_{iab} S_j{}^{ab} = 0 \tag{21}$$

we learn that $R=S_{iab}S^{iab}$ so that c(2 loops) = $a' R(X^i_o)$; R is the curvature scalar from $g^{\mu\nu}$ alone. Now a' must be the same for all manifolds, so it can be determined by calculating c(2 loops) for any special manifold ; in particular, one may pick out the two-loop coefficient in c(G) (17) for group manifolds where c_v = R/d. Recalling that the expansion is in terms of the coupling $\lambda^2 = 4\pi/K$ and using

$$c(G) = d\left(1 - \frac{\lambda^2}{4\pi}\frac{R}{d} + \left(\frac{\lambda^2}{4\pi}\frac{R}{d} \right)^2 - \dots \right) \qquad (22)$$

we deduce

$$c(2 \text{ loops}) = - \frac{\lambda^2}{4\pi} R(X\ ) \qquad \text{for all } M. \qquad (23)$$

Thus we discover th at if c(M) is to be a constant the scalar curvature $R(X^i_0)$ must be a constant function of coordinates. Otherwise, c(M) is a function of the background fields X^i_0 , which means c(M) is an operator : We integrate over constant fields X^i_0 in the path integral; consequently functions of $c(X^i_0)$ cannot be treated as simple numbers and taken outside of expectation values.

If we now use what we learn from c(2 loops), namely that $R(X^i_0)$ must be a constant; and if we once again use the special case (17) of c(G) for group manifolds, picking out the three-loop (order λ^4) term in (22), then we can simplify c(3 loops) :

$$c(3 \text{ loops}) = b\, R_{ij} R^{ij} + \left\{ \left(\frac{\lambda^2}{4\pi} \right)^2 - b \right\} R_{abcd} R^{abcd} + c\,(\nabla^2 S_{abc}) S^{abc} \qquad (24)$$

although to determine the remaining coefficients b and c, one must calculate the three-loop graphs. This result clearly agrees with the K^{-2} term in the expansion (22) of c(G) since for group manifolds $R_{ij} R^{ij} = R_{abcd} R^{abcd} = R^2/d$ and $\nabla^2 S_{abc} = 0$.

A similar analysis for the supersymmetric extension of the model with torsion (12),(13), shows that the fermions contribute the <u>additional</u> amount c_F to c(M) -- to two-loops --

<u>Supersymmetric model with torsion: general R</u>

$$c_F(1 \text{ loop}) = d/2$$

$$c_F(2 \text{ loops}) = a R + b S_{abc} S^{abc} \quad . \qquad (25)$$

Using the exact result for the supersymmetric model for group manifolds[11], $c_F = d/2$, together with the generalized Ricci flatness condition, Table I, and equation (21) we deduce

$$c\ (2 \text{ loops}) = a' \mathcal{R} \quad = 0$$

with $\mathcal{R}$ the generalized scalar curvature. Clearly if $\mathcal{R}_{abcd} = 0$, one expects $c_F = d/2$ to all orders. However, since in the supersymmetric case we only require the weaker condition $\mathcal{R}_{ij} = 0$, there exists the possibility that a $\mathcal{R}_{abcd}\mathcal{R}^{abcd}$ term may appear at order λ^4 (3 loops).

Discussion (models with torsion)

We have found that for string models with torsion the central charge c(M) is not a numerical constant unless the scalar curvature R of the manifold (from $g^{\mu\nu}$ alone) is a constant : for string theories based on the models discussed here --those with only kinetic and torsion terms-- R must be a constant. We also know from Table I -- the conditions for conformal symmetry-- that the manifoldd must have $\mathcal{R}_{abcd} = 0$ in the bosonic case and $\mathcal{R}_{ij} = 0$ in the supersymmetric case. Are there manifolds which satisfy the conditions of Table I but do not have R constant? That is, does the restriction to constant R rule out manifolds which are otherwise acceptable?

In the bosonic case, the very severe requirement that the manifold be parallelizable ($\mathcal{R}_{abcd} = 0$) does appear to exclude manifolds with R constant since the only known parallelizable manifolds are group manifolds, with one exception (see reference 2), which already have R constant. However, in the supersymmetric case , we require only the weaker condition $\mathcal{R}_{ij}=0$. Manifolds which are Ricci flat in the generalized sense, but not flat, if they exist, would likely not have R=constant, although to my knowledge none are known.

Therefore, the restriction R=constant does not eliminate any known manifolds which satisfy the conditions of Table I, since the only such known manifolds are homogeneous spaces --locally group manifolds-- which already have R constant. However, we are able, by a very simple approximate calculation of the central charge c(M), to eliminate the other possibilities without appealing to theorems about manifold parallelizability and without trying to construct manifolds with $\mathcal{R}_{ij}=0$, $\mathcal{R}_{abcd}\neq 0$, or prove that none exist.

Supersymmetric Model Without Torsion

Another interesting model is the supersymmetric extension of (12) without torsion ($e_{\mu\nu}=0$). For this model, Table I tells us we must have $R_{ij}=0$. Using this Ricci flatness condition, one may perform a loop expansion, with the result --to three loops --

$$c(1\ \text{loop}) = \tfrac{3}{2}\, d \qquad \text{Supersymmetric Model}$$

$$c(2\ \text{loops}) = 0 \qquad \text{No Torsion: } R_{ij}=0$$

$$c(3\ \text{loops}) = a\, R_{abcd}R^{abcd} \tag{26}$$

Since it is known that there exist no Ricci flat manifolds for which $R_{abcd}R^{abcd}$ is a constant, we learn that either the coefficient a in c(3 loops) must vanish or the central charge c(M) can never be a constant.

To determine the coefficient, a, requires a rather lengthy three loop claculation ; to avoid doing the calculation, I have developed a far from rigorous argument for believing a=0:

It has been proven[7] that the central charge c(M)

must be a numerical constant if the non-linear σ model, before coupling to two-dimensional gravity,

1) is conformally invariant; the β function(s) vanish

2) has a positive energy spectrum

3) has a non-degenerate ground state.

In the supersymmetric model (without torsion) 1) is satisfied if the manifold is Ricci flat and we expect 2) to be satisfied because of supersymmetry. Also, in two-dimensions only discrete symmetries can be spontaneously broken, therefore we expect 3) to be satisfied as long as the manifold M is simply connected. However, there *exist* Ricci flat manifolds which *are* simply connected (although not all are). Therefore, for *such* manifolds we expect all three criteria to be satisfied, and the coefficient a in (26) must vanish. But a is a universal coefficient -- the same for all M -- so it must be zero for all M.

Dynamics versus Geometry (models with torsion)

For the models with torsion, a similar argument

offers the intriguing possibility that one may learn purely geometrical facts about Riemannian manifolds by knowing dynamical facts about non-linear σ models. This would happen for example if one could establish dynamically that the spectrum for the supersymmetric model with torsion must have positive energy and a non-degenerate ground state, for then one would be forced to conclude that either 1) no manifolds exist for which $R_{ij}=0$ but $R_{abcd}\neq 0$ or 2) all manifolds with $R_{ij}=0$ have R = constant.

Ackowledgements

I thank the organizers of the Lewes Workshop on Superstrings, L. Clavelli and A. Halprin, for the opportunity to give this talk. This work was supported in part by the NSF under grant no. 82-01094.

References

1. A.M.Polyakov, Phys. Lett. 103B, 207, 211 (1981).

2. T.L.Curtright and C.K.Zachos, Phys. Rev. Lett.53, 1799 (1984); and with E. Braaten, University of Florida preprint #UFTP-85-01.

3. L.Alvarez-Gaume, D.Z.Freedman, and S.Mukhi, Ann. of Phys. 134, 85 (1981).

4. C.M.Hull, MIT preprint; L.Alvarez-Gaume and P.Ginsparg, Harvard University preprint; K.Stelle and P. Townsend.

5. D.Nemenschansky and S.Yankielowicz,Phys.Rev.Lett. 54 (1985).

6. S.Jain, R. Shankar, S.R.Wadia, Tata Institute (Bombay) preprint #TIRF/TH/85-3 (1985).

7. I.T.Todorov, Phys. Lett. 153B, 77 (1985);Bulg. J. Phys. 12 (1985) No.1; ISAS lecture notes 84/E.P. (Trieste).

8. V.G.Knizhnik and A.B.Zamolodchikov, Nucl. Phys. B247, 83 (1984).

9. A.N.Redlich, Brandeis preprint #BRX-TH-186 (1985).

10. D.Friedan, Phys. Rev. Lett. 45, 1057 (1980); Thesis, University of California, Berkeley (198) (to be published).

11. V.G.Kac and I.T.Todorov, MIT Math Dept. preprint (1985)

Reparametrizations and Gauge and General-Coordinate Transformations in String Field Theory

SUMIT R. DAS[†]

and

MARK A. RUBIN

Fermi National Accelerator Laboratory

Batavia, Illinois 60510

ABSTRACT

We relate reparametrizations of the parameter σ to point transformations of scalar fields in "loop space," the configuration space of string field theory. Formulas are given for the changes induced by these transformations on the infinite set of "component" spacetime-tensor fields into which a scalar field on loop space may be decomposed. New derivative operators on loop space are defined, motivated by the parametrization-dependence of the mapping from loop space to spacetime. A generalization to loop space of the Einstein-Hilbert Lagrangian is proposed as a candidate for a 2nd-quantized string Lagrangian not tied to any preferred background geometry.

†Address after November 1: Department of Physics, California Institute of Technology, Pasadena, CA 91125

I. MOTIVATIONS FOR CONSTRUCTING A STRING FIELD THEORY

In describing the quantized dynamics of point particles, we can utilize either a 1st-quantized or a 2nd -quantized formalism ("field theory"). Until recently, work on the dynamics of strings has, with few exceptions [1-3], made exclusive use of 1st quantization. First-quantized methods are well-suited for studying scattering problems, since the propagators of a 1st-quantized theory describe small fluctuations of fields about their background values.

It is in addressing questions involving <u>large</u> fluctuations of the background fields that 2nd quantization is generally of value. The most familiar example of a "large" fluctuation playing an important role in physics is the Higgs mechanism [10]. Another example, of less certain (to date) <u>physical</u> relevance, but of supreme <u>theoretical</u> relevance for string theories, is the spontaneous compactification of spatial dimensions in a theory which is defined with more than three of them to begin with (such as every string theory currently believed to be consistent). In the process of spontaneous compactification it is the gravitational field whose background value is of relevance. Since the gravitational field is one of the "component" dynamical degrees of freedom of a string theory, we must be able to deal with "large fluctuation questions" in string theories if we are to understand how such theories, constrained as they are to twenty-six or ten dimensions, somehow yield our (apparently) four-dimensional world. (Besides gauge-symmetry breaking and spontaneous compatification, there are, of course, other interesting--if perhaps less crucial--questions which 1st-quantized methods seem ill-suited to tackle, e.g.: Do "stringy" black holes have singularities? Do initial singularities occur in "stringy" cosmologies? What is the spectrum of primordial fluctuations in these cosmologies?...)

If we are to construct a string field theory to deal with problems of this nature, it is thus clear that the string field theory we construct should have the following property: The background values of all fields, especially the gravitational field, should be determined dynamically, without any one set of background values holding a preferential position in the very formulation of the theory. Otherwise we may simply be putting in by hand that which should properly be the theory's output.

Unfortunately this property is not possessed by any candidate string field theory with which the present authors are familiar [1-9]. In particular, these theories are constructed in terms of a flat background geometry, and the dynamical gravitational degrees of freedom are deviations from this preferred non-dynamical background. So, issues of fundamental interpretation aside, these theories are technically best suited for studying geometrical questions involving small fluctuations away from flat spacetime; precisely the case where 1st-quantized theory is applicable.

Needless to say, what is required is a 2nd quantized string action not tied to a special background geometry. This action must contain symmetries which, in the particle-field-theory ("zero-slope") limit, reduce to the gauge and general-coordinate invariances which 1st-quantized string theories possess in this limit. One possible source of symmetries is the geometry of loop space. In the following two sections we will study loop space and fields living on loop space, investigating those properties which are independent of the metric of the spacetime from which the loop space is constructed, or of a choice of action for the fields. In the final section we will discuss the relation of loop-space transformations to physical gauge symmetries, and will propose a candidate for a string field theory action formulated independently of a preferred background geometry.

II. REPARAMETRIZATIONS AS TRANSFORMATIONS ON LOOP SPACE

In particle field theory, the dynamical variables are fields which are functions of zero-dimensional subsets of spacetime, i.e., points. In string field theory the fields are functions of one-dimensional subsets of spacetime, i.e., curves $x^{\mu}(\sigma)$, $\mu=1,\ldots,d$, $0\leq\sigma\leq\pi$. Strictly speaking, the fields are functionals of $x^{\mu}(\sigma)$, since it takes a d-fold infinity of numbers to specify the curve $x^{\mu}(\sigma)$. A functional may be regarded as a limiting case of a function of a finite number of variables:

$$\Psi[x^\mu(\sigma)] \sim \Psi[x^\mu(0.01), x^\mu(0.02), \ldots, x^\mu(\pi)] \tag{2.1}$$

A function of many variables will, in general, have a different value if we change the order of its arguments. For example, say $x^\mu(\sigma)$ describes a closed curve, $x^\mu(\pi)=x^\mu(0)$. Let

$$\sigma \to \bar{\sigma} = \sigma - 0.02 \tag{2.2}$$

and define

$$\bar{x}^\mu(\sigma) \equiv x^\mu(\bar{\sigma}) \tag{2.3a}$$

$$\bar{\Psi}[x^\mu(\sigma)] \equiv \Psi[\bar{x}^\mu(\sigma)] \tag{2.3b}$$

Then

$$\bar{\Psi}[x^\mu(\sigma)] \sim \Psi[x^\mu(\pi-0.01), x^\mu(\pi), \ldots x^\mu(\pi-0.02)] \tag{2.4}$$

and if the initial function Ψ is chosen arbitrarily, it will in general be true that

$$\bar{\Psi}[x^\mu(\sigma)] \neq \Psi[x^\mu(\sigma)]$$

Thus, the space in which the functionals $\Psi[x^\mu(\sigma)]$ live is the space of <u>parametrized</u> <u>curves</u> in d-dimensional spacetime. We will refer to this space as "loop space". (For simplicity we restrict our attention to closed curves, except where indicated otherwise).

An infinitesimal motion of a point in loop space corresponds to an infinitesimal displacement in spacetime of each point of the curve $x^\mu(\sigma)$ to yield the curve $\bar{x}^\mu(\sigma)$;

$$\bar{x}^\mu(\sigma) = x^\mu(\sigma) + \varepsilon v^\mu(\sigma) \tag{2.5}$$

where ε is an infinitesimal parameter and $v^\mu(\sigma)$ is a spacetime vector at the point on the curve x^μ with parameter label σ. So, a vector in loop space at the loop-space point $x^\mu(\sigma)$ is a vector field in spacetime defined along the parametrized curve in spacetime $x^\mu(\sigma)$. To each parametrized curve in spacetime we can associate a different spacetime vector field, thus obtaining a loop space vector field $v^\mu[x](\sigma)$.

We can now identify certain point tranformations in loop space, $x^\mu(\sigma)\to\bar{x}^\mu(\sigma)$, with "reparametrizations" of σ, $\sigma\to\bar{\sigma}$. A point transformation in loop space is a reparametrization if it maps each point $x^\mu(\sigma)$ on the original curve to a point $\bar{x}^\mu(\sigma)$ which is also a point of the original curve $x^\mu(\sigma)$. I.e., $x^\mu(\sigma)\to\bar{x}^\mu(\sigma)$ is a reparametrization if there exists a function $\bar{\sigma}(\sigma)$ such that, for all σ,

$$\bar{x}^\mu(\sigma) = x^\mu(\bar{\sigma}(\sigma)) \ . \tag{2.6}$$

Since we can define reparametrizations which differ from one curve to another, we should really write $\bar{\sigma}[x](\sigma)$; or, for infinitesimal reparametrizations,

$$\bar{\sigma} = \sigma + \varepsilon\, g[x](\sigma) \ . \tag{2.7}$$

Consider an infinitesimal point transformation of the form (2.5); if this transformation is an infinitesimal reparametrization, then, using (2.7),

$$\begin{aligned} x^\mu(\sigma) + \varepsilon v^\mu[x](\sigma) &= x^\mu(\sigma+\varepsilon g[x](\sigma)) \\ &\approx x^\mu(\sigma)+\varepsilon g[x](\sigma)\,\frac{dx^\mu}{d\sigma} \ . \end{aligned} \tag{2.8}$$

So, if a vector field in loop space is to correspond to an infinitesimal reparametrization, it must be of the form

$$v^\mu_R[x](\sigma) = g[x](\sigma)x'^\mu(\sigma) \tag{2.9}$$

where

$$x'^{\mu}(\sigma) \equiv \frac{dx^{\mu}}{d\sigma} \quad . \tag{2.10}$$

That is, the corresponding spacetime vector field $v^{\mu}[x](\sigma)$ must be everywhere tangent to the curve $x^{\mu}(\sigma)$, as we might intuitively expect.

Having identified point transformations in loop space corresponding to infinitesimal σ reparametrizations, we are now in a position to define the action of reparametrizations on scalar functions on loop space. If $\Psi[x]$ is a scalar functional of the curve $x^{\mu}(\sigma)$, then the "reparametrized" functional $\bar{\Psi}[x]$ is a scalar functional which has the same value at $x^{\mu}(\sigma)$as $\Psi[x]$ has at $\bar{x}^{\mu}(\sigma)=x^{\mu}(\sigma)+\epsilon v_R^{\mu}[x](\sigma)$:

$$\bar{\Psi}[x^{\mu}] \equiv \Psi[\bar{x}^{\mu}] = \Psi[x^{\mu}+\epsilon v_R^{\mu}[x]] \tag{2.11}$$

To $O(\epsilon)$,

$$\Psi[\bar{x}\ ]^{\mu} = \Psi[x] + \epsilon\, \Delta_{v_R}\Psi[x] \tag{2.12}$$

where the derivative operator Δ_{v_R} [11],

$$\Delta_{v_R}\Psi[x] \equiv \int_0^{\pi} d\sigma v_R^{\mu}[x](\sigma)\, \frac{\delta\Psi[x]}{\delta x^{\mu}(\sigma)} \tag{2.13}$$

can of course be defined for any loop-space vector field $v^{\mu}[x](\sigma)$, not just those of the form (2.9).

III. REPARAMETRIZATIONS AS TRANSFORMATIONS OF TENSOR FIELDS ON SPACETIME

To investigate the relation between infinitesimal reparametrizations and linearized gauge transformations we must first describe the relation between $\Psi[x(\sigma)]$ and its "component" tensor fields [1,2]. Expand $\Psi[x(\sigma)]$ in a Taylor series about a point in loop space which also happens to be a point in spacetime; i.e., the zero-length curve

$$\overset{\circ}{x}^{\mu} \equiv \int_0^{\pi} d\sigma x^{\mu}(\sigma) \quad . \tag{3.1}$$

Then

$$\Psi[x^{\mu}] = \Psi[\overset{\circ}{x}^{\mu}] + \int_0^{\pi} d\sigma \left.\frac{\delta\Psi[x]}{\delta x^{\mu}(\sigma)}\right|_{x^{\mu}=\overset{\circ}{x}^{\mu}} (x^{\mu}(\sigma)-\overset{\circ}{x}^{\mu}) \tag{3.2}$$

$$+ \frac{1}{2!} \int_0^{\pi} d\sigma_1 d\sigma_2 \left.\frac{\delta^2\Psi[x]}{\delta x^{\mu_1}(\sigma_1)\delta x^{\mu_2}(\sigma_2)}\right|_{x^{\mu}=\overset{\circ}{x}^{\mu}} (x^{\mu_1}(\sigma_1)-\overset{\circ}{x}^{\mu_1})(x^{\mu_2}(\sigma_2)-\overset{\circ}{x}^{\mu_2}) + \ldots$$

Choose as a basis for functions of σ, $0\le\sigma\le\pi$, a set of functions satisfying

$$\int_0^{\pi} d\sigma f^{\dagger}_{\ell}(\sigma) f_m(\sigma) = \delta_{\ell m} \tag{3.3a}$$

$$\sum_{\ell} f_{\ell}(\sigma_1) f_{\ell}(\sigma_2) = \delta(\sigma_1-\sigma_2) \tag{3.3b}$$

$$f_0(\sigma) = f^{\dagger}_0(\sigma) = \frac{1}{\sqrt{\pi}} \quad . \tag{3.3c}$$

Then $\Psi\,[x^{\mu}(\sigma)]$ may be expressed in terms of an infinite number of tensor fields which live on the spacetime manifold with coordinates $\overset{\circ}{x}^{\mu}$:

$$\Psi[x^{\mu}(\sigma)] = \sum_{I=0}^{\infty} \sum_{\ell_1,\ldots\ell_I\neq 0} x^{\mu_1}_{\ell_1} \ldots x^{\mu_I}_{\ell_I} B^{\ell_1\ldots\ell_I}_{\mu_1\ldots\mu_I}(\overset{\circ}{x}) \tag{3.4}$$

where (for $\ell\neq 0$)

$$B^{\ell_1\cdots\ell_I}_{\mu_1\cdots\mu_I}(\overset{\circ}{x})$$

$$= \frac{1}{I!}\int_0^\pi d\sigma_1\ldots d\sigma_I\left\{\frac{\delta}{\delta x^{\mu_1}(\sigma_2)}\cdots\frac{\delta}{\delta x^{\mu_I}(\sigma_I)}\Psi[x]\right\}\bigg|_{x=\overset{\circ}{x}} f_{\ell_1}(\sigma_1)\ldots f_{\ell_1}(\sigma_I) \quad (3.5)$$

and

$$x^\mu_\ell \equiv \int_0^\pi d\sigma f^\dagger_\ell(\sigma)x^\mu(\sigma) \quad . \quad (3.6)$$

We now study how the components

$$B^{\ell_1\cdots\ell_I}_{\mu_1\cdots\mu_I}$$

of Ψ change under an infinitesimal reparametrization of the form (2.8). We will restrict our attention to curve-independent reparametrizations; that is, reparametrizations generated by loop-space vector fields of the form

$$v^\mu_r(\sigma) = g(\sigma)x'^\mu(\sigma) \quad (3.7)$$

rather than the more general form (2.8) (Operators of this form appear as parts of the Virasoro operators of first-quantized string theory, as we discuss in section IV. Reparametrization fields of the more general form (2.8) may be of importance, but we shall not consider them here).

Define

$$g_\ell \equiv \int_0^\pi d\sigma f^\dagger_\ell(\sigma)g(\sigma) \quad (3.8a)$$

so

$$g(\sigma) = \sum_\ell g_\ell f_\ell(\sigma) \quad . \quad (3.8b)$$

Using the above definition, (2.13), (3.4),(3.7), and the relations

$$\frac{\delta x^\mu_\ell}{\delta x^\nu(\sigma)} = \delta^\mu_\nu f^\dagger_\ell(\sigma) \ , \qquad \ell \neq 0 \tag{3.9a}$$

$$\frac{\delta x^\mu}{\delta x^\nu(\sigma)} = \delta^\mu_\nu \frac{1}{\pi} \tag{3.9b}$$

which follow from (3.1) and (3.6), we find that

$$\Delta_{V_R}\Psi[x] = \sum_{I=1} \sum_{\ell_1 \ldots \ell_I \neq 0} x^{\mu_1}_{\ell_1} \ldots x^{\mu_I}_{\ell_I} \tag{3.10}$$

$$\cdot \left\{ \frac{2i\ell_1}{\pi} g_{-\ell_1} \frac{\partial}{\partial \mathring{x}^{\mu_1}} B^{\ell_2 \ldots \ell_I}_{\mu_2 \ldots \mu_I}(\mathring{x}) + \sum_b g_b \frac{2iI\ell_1}{\sqrt{\pi}} \right.$$

$$\left. \cdot (1-\delta_{\ell_1,-b}) B^{\ell_1+b_1 \ldots \ell_I}_{\mu_1 \ldots \mu_I}(\mathring{x}) \right\}$$

In (3.10) we have chosen

$$f_\ell(\sigma) = \frac{1}{\sqrt{\pi}} e^{2i\ell\sigma} \tag{3.11}$$

and used the relation

$$B^{\ell_1 \ldots \ell_a \ldots \ell_b \ldots \ell_I}_{\mu_1 \ldots \mu_a \ldots \mu_b \ldots \mu_I} = B^{\ell_1 \ldots \ell_b \ldots \ell_a \ldots \ell_I}_{\mu_1 \ldots \mu_b \ldots \mu_a \ldots \mu_I}$$

which follows from the definition (3.5).

The components of any scalar function on loop space may be obtained using (3.5) or, equivalently, by taking derivatives with respect to the x^μ_ℓ's and evaluating the result at $x^\mu_\ell = 0$ (i.e., at $x^\mu(\sigma) = \mathring{x}^\mu$). The first three components of (2.12) are, therefore

$$(\bar{\Psi} - \Psi - \varepsilon \, \Delta_{v_r} \Psi)\Big|_{x^\mu_\ell = 0} = 0 \tag{3.12}$$

$$\frac{\partial}{\partial x^\alpha_m} (\bar{\Psi} - \Psi - \varepsilon \, \Delta_{v_r} \Psi)\Big|_{x^\mu_\ell = 0} = 0 \tag{3.13}$$

$$\frac{\partial}{\partial x^{\alpha_1}_{m_1}} \frac{\partial}{\partial x^{\alpha_2}_{m_2}} (\bar{\Psi} - \Psi - \varepsilon \, \Delta_{v_r} \Psi)\Big|_{x^\mu_\ell = 0} = 0 \tag{3.14}$$

Using (3.4) and (3.10) and defining

$$\delta B^{m_1 \ldots m_I}_{\alpha_1 \ldots \alpha_I} (\mathring{x}) = \bar{B}^{m_1 \ldots m_I}_{\alpha_1 \ldots \alpha_I} (\mathring{x}) - B^{m_1 \ldots m_I}_{\alpha_1 \ldots \alpha_I} (\mathring{x}) \tag{3.15}$$

where $\bar{B}^{m_1 \ldots m_I}_{\alpha_1 \ldots \alpha_I}$ ($B^{m_1, \ldots m_I}_{\alpha_1 \ldots \alpha_I}$) is a component of $\bar{\Psi}(\Psi)$, (3.12)-(3.14) become

$$\delta B(\mathring{x}) = 0 \tag{3.16}$$

$$\delta B^\alpha_m(\mathring{x}) = \frac{2im\varepsilon}{\sqrt{\pi}} \left\{ \frac{g_{-m}}{\sqrt{\pi}} \frac{\partial B(\mathring{x})}{\partial \mathring{x}^\alpha} + \sum_b g_b (1-\delta_{m,-b}) B^{m+b}_\alpha (\mathring{x}) \right\} \tag{3.17}$$

$$\begin{aligned} \delta B^{\alpha_1 \alpha_2}_{m_1 m_2} (\mathring{x}) = {} & \frac{4i\varepsilon}{\sqrt{\pi}} (1-\delta_{\mu_{m_1},0})(1-\delta_{\mu_{m_2},0}) \\ & \cdot \Big\{ m_1 \Big[\frac{g_{-m_1}}{2\sqrt{\pi}} \frac{\partial}{\partial \mathring{x}^{\alpha_1}} B^{m_2}_{\alpha_2}(\mathring{x}) + \sum_b g_b (1-\delta_{m_1,-b}) B^{m_1+b, m_2}_{\alpha_1 \quad \alpha_2} (\mathring{x}) \Big] \\ & + m_2 \Big[\frac{g_{-m_2}}{2\sqrt{\pi}} \frac{\partial}{\partial \mathring{x}^{\alpha_2}} B^{m_1}_{\alpha_1}(\mathring{x}) + \sum_b g_b (1-\delta_{m_2,-b}) B^{m_2+b, m_1}_{\alpha_2 \quad \alpha_1} (\mathring{x}) \Big] \Big\} \end{aligned} \tag{3.18}$$

In (3.16)-(3.18), we have computed the changes in the components

$$B^{\ell_1 \cdots \ell_I}_{\mu_1 \cdots \mu_I}(\overset{\circ}{x})$$

of Ψ with regard to a fixed basis

$$x^{\ell_1}_{\ell_1} \cdots x^{\mu_I}_{\ell_I} \; ;$$

schematically,

$$\Delta \Psi \sim \delta B^{\ell}_{\mu} \cdot x^{\mu}_{\ell} \quad . \tag{3.19}$$

There are other ways to define the effect of a derivative operator on the component fields. In general, the x^{μ}_{ℓ}'s are changed under reparametrization. If

$$\bar{\sigma} = \sigma + \varepsilon g(\sigma) \tag{3.20}$$

and

$$\bar{x}^{\mu}(\sigma) = x^{\mu}(\sigma) = x^{\mu}(\sigma) \rightarrow \varepsilon g(\sigma) x'^{\mu}(\sigma) \tag{3.21}$$

then

$$\bar{x}^{\mu}_{\ell} = x^{\mu}_{\ell} + \varepsilon \int_0^{\pi} d\sigma f^{\dagger}_{\ell}(\sigma) g(\sigma) x'^{\mu}(\sigma) \tag{3.22}$$

(In obtaining (3.22) we require that the functions $f_{\ell}(\sigma)$ retain their form under reparametrization, i.e.,

$$\bar{f}_{\ell}(\bar{\sigma}) = f_{\ell}(\bar{\sigma}) \tag{3.23}$$

To do otherwise would be to introduce at the outset a distinction between different choices of parametrization). Even the spacetime coordinates $\overset{\circ}{x}^{\mu}$ are not invariant under arbitrary reparametrizations:

$$\bar{\overset{\circ}{x}}^{\mu} = x^{\mu} + \frac{\varepsilon}{\pi}\int_0^{\pi} d\sigma g(\sigma) x'^{\mu}(\sigma) \quad . \tag{3.24}$$

For any reparametrization (3.20), (3.21), define the derivative operator $\mathscr{D}_g$ by the relations

$$\mathscr{D}_g\, x^{\mu}_{\ell} = \lim_{\varepsilon\to 0}\frac{1}{\varepsilon}(\bar{x}^{\mu}_{\ell} - x^{\mu}_{\ell}) \ , \ \ell \neq 0 \tag{3.25a}$$

$$\mathscr{D}_g\, \overset{\circ}{x}^{\mu}_{\ell} = \lim_{\varepsilon\to 0}\frac{1}{\varepsilon}(\bar{\overset{\circ}{x}}^{\mu}_{\ell} - \overset{\circ}{x}^{\mu}_{\ell}) \ . \tag{3.25a}$$

Using (3.23)-(3.25) we find that

$$\mathscr{D}_g x^{\mu}_{\ell} = \sum_m x^{\mu}_m [\frac{2im}{\sqrt{\pi}} g_{\ell-m}] \ , \quad \ell \neq 0 \tag{3.26}$$

$$\mathscr{D}_g \overset{\circ}{x}^{\mu}_{\ell} = \sum_m x^{\mu}_m [\frac{2im}{\pi} g_{-m}] \quad . \tag{3.27}$$

To obtain the action of $\mathscr{D}_g$ on the components

$$B^{\ell_1 \cdots \ell_I}_{\mu_1 \cdots \mu_I}(\overset{\circ}{x}) \ ,$$

we require that $\mathscr{D}_g$ reduce to the directional derivative in the direction $v^{\mu}_r[x](\sigma) = g(\sigma)x'^{\mu}(\sigma)$ when acting on scalars,

$$\mathscr{D}_g \Psi = \Delta_{v_r} \Psi \ , \tag{3.28}$$

and that $\mathscr{D}_g$ obey the Leibniz rule; schematically,

$$\mathscr{D}_g \Psi = (\mathscr{D}_g B^{\ell}_{\mu})\cdot x^{\mu}_{\ell} + B^{\ell}_{\mu}\cdot(\mathscr{D}_g x^{\mu}_{\ell}) \quad . \tag{3.29}$$

We find that

$$B^{\ell_1\cdots\ell_I}_{\mu_1\cdots\mu_I}(\overset{\circ}{x}) = (1-\delta_{I,0})\sum_{k=1}^{I}\frac{2i\ell_k}{\pi I} g_{-\ell_k}\frac{\partial}{\partial x^{\mu_k}} B^{\ell_1\cdots\ell_{k-1}\,\ell_{k+1}\cdots\ell_I}_{\mu_1\cdots\mu_{k-1}\,\mu_{k+1}\ldots\mu_I}(\overset{\circ}{x}) , \tag{3.30}$$

for example:

$$B(\overset{\circ}{x}) = 0 \tag{3.31}$$

$$\mathscr{D}_g B^{\ell}_{\mu}(\overset{\circ}{x}) = \frac{2i\ell}{\pi} g_{-\ell}\frac{\partial B(\overset{\circ}{x})}{\partial \overset{\circ}{x}^{\mu}} \tag{3.32}$$

$$\mathscr{D}_g B^{\ell_1\ \ell_2}_{\mu_1\ \mu_2}(\overset{\circ}{x}) = \frac{i\ell_1}{\pi} g_{-\ell_1}\frac{\partial}{\partial \overset{\circ}{x}^{\mu_1}} B^{\ell_2}_{\mu_2}(x) + \frac{i\ell_2}{\pi} g_{-\ell_2}\frac{\partial}{\partial \overset{\circ}{x}^{\mu_2}} B^{\ell_1}_{\mu_1}(\overset{\circ}{x}) \tag{3.33}$$

Equation (3.33) may be reexpressed in terms of its antisymmetric and symmetric parts:

$$\mathscr{D}_g B^{\ell_1\ell_2}_{[\mu_1\mu_2]}(\overset{\circ}{x}) = \partial_{[\mu_1}\,{}^{(-)}\xi^{\ell_1\ell_2}_{\mu_2]}(\overset{\circ}{x}) \tag{3.34}$$

and

$$\mathscr{D}_g B^{\ell_1\ell_2}_{(\mu_1\mu_2)}(\overset{\circ}{x}) = \partial_{(\mu_1}\,{}^{(+)}\xi^{\ell_1\ell_2}_{\mu_2)}(\overset{\circ}{x}) \tag{3.35}$$

where

$${}^{(\pm)}\xi^{\ell_1\ell_2}_{\mu} \equiv \frac{i\ell_1}{\pi} g_{-\ell_1} B^{\ell_1}_{\mu} \pm \frac{i\ell_2}{\pi} g_{-\ell_2} B^{\ell_1}_{\mu} . \tag{3.36}$$

The transformations (3.32), (3.34) and (3.35) are of the forms of linearized gauge transformations of Maxwell, Kalb-Ramond and gravitational fields, respectively.

(In a physical closed-string field the vector component is removed upon imposing--either by restricting the space of states or by a projection operator in the Lagrangian--the requirement that Ψ be invariant under <u>uniform</u> curve-independent reparametrizations; i.e.,

$$\Delta_{v_0}\Psi = 0 \tag{3.37}$$

where

$$v_0^\mu[x](\sigma) = g_0 x'^\mu(\sigma), \quad g_0 = \text{constant} \quad . \tag{3.38}$$

The reader can verify that (3.37) implies

$$B^{\ell_1 \dots \ell_I}_{\mu_1 \dots \mu_I}(\mathring{x}) = 0 \quad \text{unless } \ell_1 + \dots + \ell_I = 0 \tag{3.39}$$

so, in particular, $B^\ell_\mu(\mathring{x}) = 0$ since $\ell \neq 0$.

For open strings the restriction (3.39) does not arise, since, for these, $g(\sigma) = g_0 =$ constant is not a reparametrization, except for $g_0 = 0$. For open strings

$$\bar{x}^\mu(0) = x^\mu(0) \quad , \quad \bar{x}^\mu(\pi) = x^\mu(\pi) \tag{3.40}$$

so (2.8),(2.9), and (3.40) imply that

$$g(0) = g(\pi) = 0 \quad . \tag{3.41}$$

We cannot invoke here the 1st-quantized equations of motion, $x'^\mu(0)=x'^\mu(\pi)=0$, since we are dealing with the field theory and $x^\mu(\sigma)$ is, not a dynamical variable, but an element of the set which indexes the dynamical variables Ψ. So, we see that the only uniform reparametrization of open curves consistent with (3.40) is the identity, $g(\sigma) = 0$).

What is the infinitesimal transformation of Ψ such that the <u>components</u> of this transformation (i.e., with respect to fixed x^μ_ℓ's) are of the form (3.30)? That is, what operator $\mathbf{D}_g$ satisfies

$$\varepsilon\, \mathbf{D}_g \Psi = (\varepsilon \mathscr{D}_g B^\ell_\mu) \cdot x^\mu_\ell \; ? \tag{3.42}$$

The required operator has the form

$$\mathbf{D}_g = \frac{2i}{\pi} \left(\sum_m m\, x^\mu{}_m g_{-m}\right) \frac{\partial}{\partial \overset{\circ}{x}{}^\mu} \quad . \tag{3.43}$$

It can be easily checked that, unless $g(\sigma)$ is identically zero, $\mathbf{D}_g$ is <u>not</u> a curve-independent reparametrization. That is, there is no function $q(\sigma)$ such that

$$\mathbf{D}_g = \Delta_{qx'} = \int_0^\pi d\sigma\, q(\sigma) x'(\sigma) \frac{\delta}{\delta x^m(\sigma)} \quad . \tag{3.44}$$

However, $\mathbf{D}_g$ does have a simple interpretation. Using (3.27b) and (3.43), we see that

$$\varepsilon\, \mathbf{D}_g = \varepsilon\, \mathscr{D}_g\, \overset{\circ}{x}{}^\mu \frac{\partial}{\partial \overset{\circ}{x}{}^\mu} = (\bar{\overset{\circ}{x}}{}^\mu - \overset{\circ}{x}{}^\mu)\, \frac{\partial}{\partial \overset{\circ}{x}{}^\mu} \quad . \tag{3.45}$$

So, $\varepsilon\, \mathbf{D}_g$ is the <u>spacetime</u> translation operator which translates by an amount proportional to the shift in spacetime coordinate $\overset{\circ}{x}{}^\mu$ associated with a reparametrization $\sigma \to \sigma + \varepsilon g(\sigma)$.

IV. DISCUSSION

Is either the Δ or $\mathbf{D}$ transformation a viable candidate to yield linearized gauge transformations in a physical string field theory? That is, should one attempt to construct loop space actions, functionals of $\Psi[x^\mu(\sigma)]$, which, at least linearly, are invariant under

$$\Psi \to \Psi + \varepsilon \Delta_{v_r} \Psi \tag{4.1}$$

or

$$\Psi \to \Psi + \varepsilon \mathbf{D}_g \Psi \; ? \tag{4.2}$$

Invariance under (4.1) or (4.2) corresponds to invariance under component-field transformations such as (3.16)-(3.18) or (3.31)-(3.35). Although, as has been noted, (3.32),(3.34), and (3.35) are similar in appearance to familiar gauge-transformation laws, both (4.1) and (4.2) differ from the usual gauge transformations in an important respect; the transformation parameters on the right-hand sides are themselves proportional to the fields which are being transformed, rather than totally arbitrary functions. Theories possessing invariance under (4.1) or (4.2) would thus be very different from conventional gauge theories [12]. If we want the usual electromagnetic, gravitational, etc., linearized gauge invariances to be contained in our theory, we must replace (4.1) and (4.2) with

$$\Psi \to \Psi + \varepsilon \Delta_{v_r} \Omega \tag{4.3}$$

$$\Psi \to \Psi + \varepsilon \mathbf{D}_g \Omega \tag{4.4}$$

where Ω is an arbitary scalar function on loop space. (Admittedly, in doing so we lose the purely geometric interpretation which (4.1), (4.2) have, although the operators Δ_{v_r} and $\mathbf{D}_g$ are still associated with geometric transformations in loop space). The equations for the components of tranformations (4.3) and (4.4) are identical to those for the components of (4.1) and (4.2), except that, on the right-hand sides,

$$B^{\ell_1 \cdots \ell_I}_{\mu_1 \cdots \mu_I}(\mathring{x})$$

is replaced by

$$\omega^{\ell_1 \cdots \ell_I}_{\mu_1 \cdots \mu_I}(\mathring{x}) \quad ,$$

where

$$\Omega[x] = \sum_{I=0}^{\infty} \sum_{\ell, \ldots \ell_I \neq 0} x^{\mu_1}_{\ell_1} \cdots x^{\mu_I}_{\ell_1} \, \omega^{\ell_1 \cdots \ell_I}_{\mu_1 \cdots \mu_I} (\mathring{x}) \quad . \tag{4.5}$$

For example, (3.32) becomes

$$B^{\ell}_{g\,\mu}(\overset{\circ}{x}) = \frac{2i\ell}{\pi}\, g_{-\ell}\, \frac{\partial\omega(\overset{\circ}{x})}{\partial \overset{\circ}{x}{}^{\mu}} \quad . \tag{4.6}$$

Even with this improvement, the rule $\Psi \to \Psi + \varepsilon D_g \Omega$ is unsuitable in another way for use as a gauge-transformation law. Under this rule, <u>all</u> the component tensor fields transform as gauge fields. So, a theory with this symmetry would have an infinite number of physical gauge fields.

As for $\Psi \to \Psi + \varepsilon \Delta_{v_r} \Omega$, comparison with first-quantized string theory suggests that this is only part of the correct rule. In the first-quantized theory in flat spacetime a crucial role is played by the Virasoro operators [13]. Each of these operators is equal to the sum of a Δ_{v_r} operator, for a suitable $g(\sigma)$, with another piece which is a Fourier transform with respect to σ of the operator

$$-\eta^{\mu\nu} \frac{\delta}{\delta x^{\mu}(\sigma)} \frac{\delta}{\delta x^{\nu}(\sigma)} + \frac{1}{4\pi^2\alpha'^2}\, \eta_{\mu\nu}\, x'^{\mu}(\sigma) x'^{\nu}(\sigma) \tag{4.7}$$

($\eta_{\mu\nu}$ is the Minkowski metric, and α' is a constant). In the context of string field theory it is not clear how to give these operators a geometrical interpretation [14]. There is of, course, a formal similarity with the mass-shell operator for point particles,

$$-\eta^{\mu\nu} \frac{\partial}{\partial x^{\mu}} \frac{\partial}{\partial x^{\nu}} + m^2 \tag{4.8}$$

in that both (4.7) and (4.8) arise from the constraint equations satisfied by the respective canonical momenta,

$$P^{\mu}(\sigma) P_{\mu}(\sigma) + \frac{1}{4\pi^2\alpha'^2}\, \eta_{\mu\nu}\, x'^{\mu}(\sigma) x'^{\nu}(\sigma) = 0 \quad , \tag{4.9}$$

$$p^{\mu} p_{\mu} + m^2 = 0 \quad ,$$

as consequences of the choice of a parametrization-invariant action for the respective classical objects [13],

$$S(\text{string}) \sim \int d\sigma\, d\tau [-(\dot{x})^2(x')^2 + (\dot{x}\cdot x')^2]^{1/2} \quad , \tag{4.11}$$

$$S(\text{particle}) \sim \int d\tau\, [-\dot{x}^2]^{1/2} \quad . \tag{4.12}$$

This analogy suggests that the Nambu action (4.11) be regarded as a distance function in loop space, and that any metric tensor on "curved" loop space should have the property that the "interval" between the loop-space points $x^\mu(\sigma)$, $y^\mu(\sigma)$ is equal, at least for "nearby" points, to the area of the minimal surface between them.

Having seen that loop-space geometry may bear at least some relation to physical gauge symmetries, we are motivated to propose a purely geometric candidate for a background-geometry-independent formulation of the string field theory action. Our proposal is a modification of an action invented by Freund and Nepomechie [15] in studying the geometry of Kalb-Ramond fields, and its potential applicability to string theory was originally noted by them. These authors take as their action the Einstein-Hilbert action constructed from a metric tensor living on a space even larger than loop space, namely, the fiber bundle (loop space) $\otimes$ U(1). After certain ad hoc restrictions are imposed on the form of the metric and the $x^\mu_\ell \to 0$ limit is taken, this action reduces to the bosonic part of the action for ten-dimensional supergravity.

Since Ψ contains objects,

$$B^{\ell m}_{(\mu\nu)} ,$$

with transformation properties similar to the linearized gauge transformation properties of a metric perturbation $h_{\mu\nu}$, it seems unwarranted to introduce an additional field in the theory to describe the geometry. Indeed, one of the beauties of string theory is precisely the fact that gravitational degrees of freedom arise from the string degrees of freedom. We therefore suggest that one take as the loop-space metric tensor, out of which to construct the loop-space Einstein-Hilbert action, the second variational derivative of the scalar string field itself (not restricted to $x^\mu_\ell = 0$!):

$$g_{\mu_1\mu_2}(\sigma_1,\sigma_2) = \frac{\delta^2\Psi[x]}{\delta x^{\mu_1}(\sigma_1)\delta x^{\mu_2}(\sigma_2)} \quad . \tag{4.13}$$

Ψ is, in general, complex, so it may not be necessary to augment loop space with an additional U(1) fiber. The results of [15] seem to indicate that such a theory will have a suitable particle-field-theory ($x^\mu_\ell \to 0$) limit; its properties as a string field theory ($x^\mu_\ell \neq 0$) are currently under investigation.

We would like to thank Michio Kaku and Stuart Raby for enlightening discussions, and the Lewes Center for Physics for warm hospitality during the course of this work.

References

1. M. Kaku and K. Kikkawa, Phys. Rev. D10 (1974)1110, 1823.
2. C. Marshall and P. Ramond, Nucl. Phys. B85 (1975) 375.
3. M. B. Green and J. H. Schwarz, Nucl. Phys. B243 (1984) 475.
4. M. Kaku, "Gauge Field Theory of Covariant Strings," CCNY preprint; "Supergauge Field Theory of Covariant Heterotic Strings," Osaka preprint OU-HET-79, July 1985. M. Kaku and J. Lykken, "Supergauge Field Theory of Covariant Superstrings," CCNY preprint; "Supergauge Field Theory of Superstrings," in Symposium on Anomalies, Geometry, Topology, World Scientific Publishing Co., Singapore, 1985.
5. T. Banks, "Gauge Invariant Functional Field Theory for Bosonic Strings," in Symposium on Anomalies, Geometry, Topology, World Scientific Publishing Co., Singapore, 1985. T. Banks and M. E. Peskin, "Gauge Invariance of String Fields," SLAC preprint SLAC-PUB-3740 July 1985.
6. A. Neveu and P. C. West, "Gauge Covariant Local Formulation of Bosonic Strings," CERN preprint CERN-TH.4200/85. A. Neveu, H. Nicolai, and P. C. West, "Gauge Covariant Local Formulation of Free Strings and Superstrings," CERN preprint CERN-TH. 4233/85
7. D. Friedan, "String Field Theory," University of Chicago preprint EFI 85-27, April 1985. This paper contains a discussion of curved-spacetime string field theory.
8. K. Bardakci, "Covariant Gauge Theory of Strings," Berkeley preprint LBL-20024, UCB-PTH-85/33, July 1985.
9. K. Itoh, T. Kugo, H. Kunitomo, and H. Oguri, "Gauge Invariant Local Action of String Field from BRS Formalism," Kyoto preprint KUNS 800, HE (TH) 85/04 August 1985.
10. For a review, see: E. S. Abers and B. W. Lee, Phys. Rep. 9C (1973)1.
11. This operator has appeared previously in ref. [2].
12. We are grateful to Stuart Raby for a helpful conversation on this point.
13. For a review, see: J. Scherk, Rev. Mod. Phys. 47, (1975) 123.

14. Nor is there a clear geometrical interpretation to the fact that, in the Lorentz-covariant string field theories of refs. [4-6], only Virasoro operators L_n with $n<0$ correspond to gauge transformations.
15. P. G. O. Freund and R. I. Nepomechie, Nucl. Phys. B199 (1982) 482.

Gauge and General Covariance of String Interactions

SUMIT R. DAS†
Fermi National Accelerator Laboratory
Batavia, Illinois 60510

I. INTRODUCTION

All fundamental interactions at observable energies seem to arise out of local symmetries - gauge invariances and general coordinate invariance. In usual field theories of point particles these invariances are postulated a priori: the idea is to deduce everything else from the symmetry group and the representation content of the matter fields.

In string theories[1], the situation is rather different. Here the basic principle is reparametrization invariance on the world sheet swept out by the string. We shall consider the simplest string models-those defined on flat Minkowski space-time. The transverse oscillations of the string lead to an infinite tower of modes which may be thought of as the "particles" constituting the string. The interacting string theory is defined, in the first quantized formulation, by specifying the interaction of these modes with the string. These interaction vertices must satisfy a basic requirement: when any dual amplitude is factorized only physical states (i.e. those satisfying the Virasoro conditions) must occur as on-mass-shell intermediate states.[1] This means that the vertices respect the reparametrization invariance of the world sheet, since it is this symmetry which eliminates ghost states by virtue of Virasoro conditions.

All massive modes of the string have masses of the order of the string tension. Therefore, at energies much smaller than the string tension (the zero slope limit) only the massless modes can be excited-these are the particles of the low-energy world. The effective zero slope limit lagrangian defines the particle field theory valid at low energies. Now, the massless spectrum of interesting string models contain spin-1 and spin-2 modes. These are potential candidates for gauge bosons and gravitons. Gauge bosons and gravitons are above all gauge particles of local symmetries. Such local symmetries are not, however, postulated a priori in string theories. Gauge principles are not invoked in writing down the interaction vertices of these spin-1 and spin-2 modes with the string. The only requirement, as stated above, is the absence of ghosts. Remarkably, however, in the zero slope limit the massless vectors and tensors interact with each other and with other fields in a way consistent with gauge and general coordinate invariance.

In this talk I shall indicate how the interaction of the massless modes with the full string respect gauge and general coordinate invariance even though the latter are not postulated a priori. This will be done by studying the propagation of a string in arbitrary backgrounds containing such massless fields.

Most of the work reported here has been done in collaboration with Spenta R. Wadia. A short account is contained in Ref. 2.

I would like to emphasize that our analysis does not solve the main mystery: how does gauge and general coordinate invariance arise in string theories. To answer this question one needs to understand string field theory fully. Let me digress from our main topic and state the issues involved.

II. LOCAL INVARIANCES IN STRING FIELD THEORY

Consider the bosonic string field theory.[3] The functional operator $\Phi[x(\sigma)]$ creates a string along the curve $x^{\mu}(\sigma)$. The string coordinate $x^{\mu}(\sigma)$ may be expanded in normal coordinates. For an open string one has

$$x^{\mu}(\sigma) = x_0^{\mu} + \sum_{n=1}^{\infty} x_n^{\mu} \frac{\cos n\sigma}{n} \tag{1}$$

A general expansion for the string field functional is of the form

$$\Phi[x(\sigma)] = \phi(x_0) + A_{\mu}(x_0)H_1(x_1^{\mu}) + \ldots.. \tag{2}$$

where e. g. $H_1(x_1{}^{\mu})$ is the harmonic oscillator wave function with the n=1 mode (of type μ) excited to the first level - the rest being in the ground state. $A_{\mu}(x_0)$ is in fact the field which creates the massless vector mode of the string. Similar considerations are, of course, valid for the closed string.

In what sense can $A_{\mu}(x_0)$ represent a "photon"-a gauge particle? Suppose the full second quantized string action is invariant under some set of transformations:

$$\Phi \rightarrow \Phi + \delta\Phi$$

This transformation would induce certain transformations on the component fields. Thus, e. g.

$$A_{\mu}(x_0) \rightarrow A_{\mu}(x_0) + \delta A_{\mu}(x_0) \quad .$$

Now suppose a $\delta\Phi$ exists which imply

$$\delta A_{\mu}(x_0) = D_{\mu}\alpha(x_0)$$

(where $D_{\mu} \equiv \partial_{\mu} + i[A_{\mu},]$ when A_{μ} is a matrix), i. e. a gauge transformation. This is possible if $\delta\Phi$ is itself of the form

$$\delta\Phi = M\,\Omega\,[x(\sigma)]$$

where M is some operator in loop space and $\Omega[x(\sigma)]$ is an arbitrary functional. $\alpha(x_0)$ may be then the first term in an expansion of Ω of the form (2). $\delta\Phi$ is a transformation in loop space which becomes usual gauge transformations on the component fields. One must, of course, find similar transformations on closed string functionals which lead to general coordinate invariance.

Do such operators like M exist? This question is addressed in the talks given by Kaku[4], Raby[5] and Rubin[6] in this workshop and I shall not discuss this in any detail. The answer seems to be yes-at least for the free theory; in which case one obtains the linearized local symmetries. We have made some progress in understanding the geometrical meaning of these loop space transformations.[6] M. Rubin shall tell you about that. Not much is known for the interacting theory; in fact a gauge invariant and Lorentz invariant version of second quantized interacting string theory has not yet been formulated.

III. STRING IN BACKGROUND FIELDS

Let us return to our main concern: how does string propagation in background fields reflect gauge and general coordinate invariance? why would one like to study such a problem? One reason is that the vacuum of the second quantized string theory is probably highly non-trivial. Consider once again the string field functional $\Phi[x(\sigma)]$. A trivial fock vacuum would be characterized by

$$\langle 0|\Phi(x(\sigma))|0\rangle = 0 \quad .$$

which means $\langle A_\mu(x_0)\rangle = 0$ and $\langle h_{\mu\nu}(x_0)\rangle = 0$, where $h_{\mu\nu}(x)$ is the spin-2 "graviton" mode or the closed string. If string theory describes nature this is surely not the true vacuum. In particular, interesting string models live in ten space-time dimensions. In the ground state the extra six dimensions must curl up into a compact and possibly curved manifold. Since $h_{\mu\nu}(x)$ is the physical graviton field, a non zero curvature means

$$\langle h_{\mu\nu}(x)\rangle \neq 0$$

In other words, the vacuum contains a condensate of gravitons Similarly, standard arguments relating to the existence of chiral fermions indicates that there must be a condensate of Yang-Mills fields as well. In fact the vacuum might contain massive fields.

We shall not ask the question as to how such a vacuum might arise. To answer that, one needs a much better understanding of the full convariant second quantized string field theory. Instead, we shall assume that such a vacuum is present and ask how a string propagates in this vacuum. For this purpose a first quantized formalism is sufficient.

(a) Scalar Electrodynamics: A Toy Example

First consider electrodynamics of a scalar Klein-Gordon particle following the treatment of Feynman.[7] If the particle trajectory is denoted by $x^\mu(\tau)$ (τ being the parameter parametrizing the path) the amplitude for a free particle to go from x_1 to x_2 is given by

$$\langle x_2|x_1\rangle = \int_{-\infty}^{+\infty} K\,(x_2,\tau_2;\, x_1,\tau_1)\, e^{-1/2im^2(\tau_2-\tau_1)}\, d\tau_2 d\tau_1 \tag{3}$$

where the kernel K is given by the path integral:

$$K(x_2,\tau_2;\, x_1,\, \tau_1) = \int \mathcal{D}x(\tau)\exp(iS_0^P) \tag{4}$$

and S_0^P is the free action:[5]

$$S_0^P = -\frac{1}{2}\int d\tau\, \left(\frac{dx^\mu}{d\tau}\cdot\frac{dx_\mu}{d\tau}\right) \tag{5}$$

In (3) m is the mass of the particle. Once the free propagator is obtained from (3) and (4) the propagator in a background electromagnetic field may be calculated in a perturbation series using the standard Feynman rules. The vertex for interaction of a photon of momentum $\underline{k}$ with the particle is:

$$V(k) = e\,\frac{dx^\mu}{d\tau}\,\varepsilon_\mu(k)\exp(ik\cdot x(\tau)) \tag{6}$$

where $\varepsilon_\mu(k)$ is the polarization vector. The full amplitude is now obtained by inserting the vertex (6) at various intermediate points between τ_1 and τ_2, integrating over these points and finally integrating over all momenta. Of course, all this is equivalent to adding the interaction term

$$S_I^P = e\int_{\tau_1}^{\tau_2} d\tau\, \frac{dx^\mu}{d\tau} A_\mu(x(\tau)) \tag{7}$$

to the free action S_0^P and performing a perturbation expansion in e. In (7) $A_\mu(x)$ is the photon field

$$A_\mu(x) = \int dK e^{ikx} \varepsilon_\mu(k) \tag{8}$$

The total action $S_0^P + S_I^P$ describes a gauge invariant interaction of the particle with an external field. Under a gauge transformation:

$$A_\mu(x) \to A_\mu(x) + \partial_\mu \alpha(x)$$

the amplitude transforms as:

$$\langle x_2|x_1\rangle \to e^{i\alpha(x_2)} \langle x_2|x_1\rangle e^{-i\alpha(x_1)} \tag{9}$$

as expected from gauge invariance.

One way of coupling the scalar particle to non-Abelian fields is to make A_μ maxtrix in adjoint representation of the gauge group. The path integral is now understood as a τ-ordered product. Consider a τ-slice between τ_i and τ_{i+1}. Under a gauge transformation of the potential at i:

$$A_\mu(x_i) \to U(x_i) A_\mu(x_i) U^{-1}(x_i) + \frac{1}{i}(\partial_\mu U(x_i)) U^{-1}(x_i)$$

(where $f(x_i) \equiv f(x(\tau_i))$ for any function), the part of the Kernel belonging to the relevant τ slice transforms as follows:

$$\langle x_{i+1}|x_i\rangle \to U(x_{i+1}) \langle x_{i+1}|x_i\rangle U^{-1}(x_i) \tag{10}$$

provided $\tau_{i+1}-\tau_i \to 0$. Thus in the naive continuum limit one recovers the usual gauge transformation property of the amplitude.

Note that the above effective action was deduced from perturbation theory. However (7) is taken to be the general interaction term even for non-perturbative external fields like that of a monopole.

In a sense the above exercise is redundant. For electrodynamics the vertex function does not have to satisfy any requirement other than gauge invariance. Hence one could impose gauge invariance a priori and obtain (7) right away and that is the end of the story. We went through this exercise because we shall use basically the same method for studying string propagation in background fields-and here the implications of our results are non-trivial.

(b) Properties of String vertices

For a string, the quanta which condense to form the background are modes of the string itself. The physical states of a string $|\Psi\rangle$ are constrained by the Virasoro conditions:

$$L_n|\Psi\rangle = 0 \quad n>0 \tag{11}$$

where L_n are the usual Virasoro generators. Consider a general string diagram describing scattering of physical on-mass-shell modes. This is a world sheet with insertions of the relevant vertex operators at various points on the sheet. When this dual amplitude is factorized, the on-mass-shell projections of the intermediate states in any channel are not any more guaranteed to satisfy the Virasoro conditions (11). This means that for arbitrary vertex operators the theory contains ghosts. For the theory to make any sense the vertex operators must satisfy certain conditions which ensure the absence of ghosts. Consider a general vertex operator $V(\tau)$. The conformal spin J_A of any operator $A(\tau)$ is defined as follows:

$$[L_n, A(\tau)] = e^{in\tau}(-i\frac{\partial}{\partial\tau} + n\, J_A)A(\tau) \tag{12}$$

In the interaction picture time evolution is governed by the free Hamiltonian L_0(in the orthonormal gauge) so that (12) implies:

$$[L_n - L_0, A(0)] = nJ_A A(0) \tag{13}$$

From these definitions it may be easily shown that absence of ghosts is ensured if[8]

$$J_V = 1 \tag{14}$$

where J_v is the conformal spin of the vertex $v(\tau)$.

In the rest of the talk I shall indicate how vertices of massless modes satisfying the above condition automatically ensures gauge and general coordinate invariance, albeit in linearized form.

IV. BOSONIC STRINGS

Let us first consider the bosonic string described by $x^\mu(\sigma_0\tau)$. The action in a convariant orthonormal gauge is given by:

$$S_0^S = \frac{1}{2\pi}\int_{\tau_1}^{\tau_2} d\tau \int_0^\pi d\sigma\ \eta_{\mu\nu}\ \partial_\alpha x^\mu \partial^\alpha x_\mu \tag{15}$$

The transition amplitude between the string states $|x_1(\sigma),\tau_1\rangle$ and $|x_2(\sigma),\tau_2\rangle$ is given by:

$$\langle X_2(\sigma),\tau_2|X_1(\sigma),\tau_1\rangle = \int DX^\mu \Delta \exp\,(iS_0^S) \tag{16}$$

where Δ is a measure term which includes gauge fixing and Fadeev Popov factors.

The massless mode of an open string is a vector. The vertex function describing the interaction of this vector with the full string is given by

$$V_v(k) = g\,\frac{\partial X^\mu(0,\tau)}{\partial\tau}\,\varepsilon_\mu(k)e^{ikX(0,\tau)} \tag{17}$$

This vertex satisfies the primary requirement, eqn. (14). Here $\varepsilon_\mu(k)$ is the polarization vector of the vector particle carrying a momentum k. (17) describes the vertex when the vector hits the $\sigma = 0$ end of the string. The corresponding vertex for the $\sigma = \pi$ end is obvious.

In a way analogous to the case of scalar electrodynamics the amplitude for string propagation in a background spin-1 field is obtained by adding to S_0^S the term:

$$S_I^{(1)} = g\int_{\tau_1}^{\tau_2} d\tau \int_0^{\pi} d\sigma \, \frac{\partial x^{\mu}(\sigma,\tau)}{\partial\tau} A_{\mu}(x(\sigma,\tau))\delta(\sigma) \qquad (18)$$

and using the full action $S_0{}^s + S_I{}^{(1)}$ in the path integral (16). For non-abelian fields $A_{\mu}(x)$ is again a matrix and one obtains the usual Chan-Paton factors. From eqn. (18) it is clear that the string propagates as if there is a charge at its end which interacts with the background field in the usual gauge invariant fashion. Unlike the case of particle theories, this gauge invariance need not be postulated a priori. The vertex (17), chosen to ensure the absence of ghosts, is automatically gauge invariant. Alternatively, one could say that if the vertex is chosen to be gauge invariant absence of ghosts follows.

Now consider the interaction of a string with the massless tensor modes of the closed string. This interaction may occur anywhere along the extension of the string. The relevant vertex operator having a conformal spin equal to one is

$$\frac{\kappa}{\pi} \int_0^{\pi} d\sigma \varepsilon_{\mu\nu}(k) \frac{\partial x^{\mu}}{\partial\tau_-} \frac{\partial x^{\nu}}{\partial\tau_+} e^{ikx(\sigma,\tau)} \qquad (19)$$

where $\tau_{\pm} = \tau \pm \sigma$. For the spin-2 mode the polarization tensor $\varepsilon_{\mu\nu}(k)$ is symmetric and transverse:

$$k_{\mu}\varepsilon_{\mu\nu} = k_{\nu}\varepsilon_{\mu\nu} = 0; \; \varepsilon_{\mu\nu} = + \varepsilon_{\nu\mu}$$

For the antisymmetric tensor mode, $\varepsilon_{\mu\nu}$ is antisymmetric.

Eq. (19) may be rewritten for symmetric $\varepsilon_{\mu\nu}$ as:

$$\frac{\kappa}{\pi} \int_0^{\pi} d\sigma \, \varepsilon_{\mu\nu}(k) \partial_{\alpha} x^{\mu} \partial^{\alpha} x^{\nu} e^{ikx(\sigma,\tau)} \qquad (20)$$

If $h_{\mu\nu}(x)$ denotes a background field of spin-2 particles. The effective action is thus given by

$$S_0^S + \frac{\kappa}{\pi}\int_{\tau_1}^{\tau_2} d\tau \int_0^\pi d\sigma h_{\mu\nu}(x)\partial_\alpha x^\mu \partial^\alpha x^\nu = \frac{1}{2\pi}\int d\tau d\sigma g_{\mu\nu}(x)\partial^\mu x^\alpha \partial^\nu_\alpha x \tag{21}$$

where

$$g_{\mu\nu} = \eta_{\mu\nu} + 2\kappa h_{\mu\nu}$$

is the total background metric in which the string propagates in a manner consistent with general covariance. Since $h_{\mu\nu}$ is transverse only the transverse components of the metric are non-trivial. Note, however, the orthonormal gauge constraints are not generally covariant - to $O(h_{\mu\nu})$ they may be replaced by generally covariant expressions. In fact our results are true for weak backgrounds since we are assuming that the spectrum of states are unchanged.

The case of an antisymmetric tensor field $A_{\mu\nu}(x)$ may be dealt with in an identical fashion. The effective action is:

$$\int d\tau d\sigma\, A_{\mu\nu}(x)\partial_\alpha x^\mu \partial_\beta x^\nu \varepsilon^{\alpha\beta} \tag{22}$$

This is precisely a Wess-Zumino term for the non-linear sigma model defined by eqn. (21) - which in turn is equivalent to a torsion.[9] Under a Kalb-Ramond type gauge transformation of $A_{\mu\nu}$:

$$A_{\mu\nu}(x) \rightarrow A_{\mu\nu}(x) + \partial_\mu \Lambda_\nu(x) - \partial_\nu \Lambda_\mu(x)$$

the strip amplitude transforms as

$$\left\langle x_W(\sigma),\tau_2 | x_1(\sigma),\tau_1 \right\rangle \rightarrow \exp\left[i \oint_{x_2(\sigma)} dx^\mu \Lambda_\mu(x) \right]$$

$$\left\langle x_2 | x_1 \right\rangle \exp\left[-i \oint_{x_1(\sigma)} dx^\mu \Lambda_\mu(x) \right]$$

Eq. (21) has been widely used in the literature to discuss strings in curved spaces.[10] Here we are adopting a different attitude. The string theory we consider lives on a flat background. However, the vacuum may contain a graviton condensate. The action (21) then provides the first quantized description of a string propagating in such a vacuum.

V. SUPERSTRINGS

The above considerations may be extended to Green-Schwarz superstrings in a straightforward manner. The free Type I superstring action in the light cone gauge is given by:

$$S_0 = \int d\sigma d\tau \left[\frac{1}{2\pi} \partial_\alpha x^i \partial^\alpha x_i + \frac{i}{4\pi} \bar{S}\gamma^- \rho^\alpha \partial_\alpha S\right] \tag{23}$$

Here x^i are the transverse coordinates and S^{Aa} denotes a ten-dimensional Majorana-Weyl spinor which also transforms as a spinor under world sheet reparametrizations. The ρ^α 's denote 2 dimensional gamma matrices while γ's are space-time gamma matrices.

The interaction of an open superstring with a background spin-1 field is very similar to that of a bosonic string; we shall not discuss this case here

Consider the interaction of a superstring with the spin-2 mode of the closed string. For a closed string, one has:

$$S^{Aa} = \begin{pmatrix} S^{1a} \\ S^{2a} \end{pmatrix} = \begin{pmatrix} S^a(\tau_-) \\ S^a(\tau_+) \end{pmatrix}$$

where a is a space-time spinor index and $\tau_\pm = \tau \pm \sigma$. The spin-2 vertex is given by:[11]

$$\frac{\kappa}{\pi} \int_0^\pi d\sigma \; B^i_c(P,\tau_-) B^j_c(P,\tau_+) e^{ikx(\sigma,\tau)} \tag{24}$$

where p is the momentum and

$$\begin{aligned} B^i_c(p,\tau_-) &= \frac{\partial x^i}{\partial \tau_-} + \frac{1}{2} p^j R^{ij}(\tau_-) \\ R^{ij}(\tau_-) &= \bar{S}(\tau_-)\gamma^{ij-} S(\tau_-) \end{aligned} \tag{25}$$

The resulting effective action may be now written down:

$$S = \int d\sigma d\tau \Big\{ -\frac{1}{2\pi} g_{ij} \partial_\alpha x^i \partial^\alpha x^j + \frac{i}{4\pi} \bar{S}\gamma^- \rho^\alpha \partial_\alpha S - \frac{i\kappa}{\pi} (\partial_j h_{ik})(\partial_\alpha x^k) \bar{S}\gamma^{ij-} \rho^\alpha S - \frac{\kappa}{4\pi} (\partial_j \partial_l h_{ik}) R^{kl}(\tau_+) R^{kj}(\tau_-) \Big\} \tag{26}$$

This expression does not appear to be generally covariant. However, to $O(h_{ij})$ one can replace all ordinary derivatives by covariant derivatives ∇_j and flat space gamma matrices by their curved space counterparts, $\hat{\gamma}^i$. Furthermore, up to the same order one can make the replacements:

$$\partial_j h_{ik} \leftrightarrow \Gamma_{jik}$$

and $\partial_j \partial_\ell h_{ik} \leftrightarrow R_{ijk\ell}$

to obtain

$$S = \int d\sigma d\tau \Big\{ -\frac{1}{2\pi} g_{ij} \partial_\alpha x^i \partial^\alpha x^j + \frac{i}{4\pi} \bar{S} \gamma^- \rho^\alpha D_\alpha S - \frac{\kappa}{24\pi} R_{ijk\ell} R^{ij}(\tau_+) R^{k\ell}(\tau_-) \Big\} + O(h^2_{ij}) \tag{27}$$

where

$$D_\alpha \equiv \partial_\alpha - \kappa \hat{\gamma}^{ij} \Gamma_{ijk} \partial_\alpha x^k$$

This is the generally covariant expression for the effective action. It is instructive to rewrite (27) in terms of the Rammond-Neveu-Schwarz fields, λ^i which are spinors on the world sheet and vectors in transverse space time. Using the standard expressions relating the "old" and "new" oscillators one gets[12]

$$S = -\frac{1}{2\pi}\int d\tau d\sigma \left\{ g_{ij}\partial_\alpha x^i \partial^\alpha x^j + i\,\bar{\lambda}^i \rho^\alpha \partial_\alpha \lambda^j g_{ij}(x) \right. \tag{28}$$

$$\left. + i\,\bar{\lambda}^\ell \rho^\alpha \partial_\alpha x^i \Gamma_{\ell ij} \lambda^j + \frac{1}{8} R_{i\ell mj} \bar{\lambda}^j (1+\gamma_S)\lambda^\ell \bar{\lambda}^m (1+\gamma_S)\lambda^i \right\} + O(h_{ij}^2)$$

Here $\gamma_S \equiv \rho^0 \rho^1$. In this form the action also has a manifest supersymmetry on the world sheet. In fact (28) is precisely the supersymmetric non-linear sigma model with the fields lying on a manifold with metric g_{ij}.

We shall not go into the details of deriving the effective action in the presence of an antisymmetric tensor field. As expected, the new interaction term is precisely the supersymmetric extension of the Wess-Zumino term.

VI. THE HETEROTIC STRING

We shall now consider briefly the propagation of a heterotic string in a given background. This has been discussed in detail in Refs. 14. The free light cone gauge action is given by:

$$S = -\frac{1}{4\pi\alpha'}\int d\tau \int_0^\pi d\sigma \, [\partial_\alpha X^i \partial^\alpha X^i + \partial_\alpha X^I \partial^\alpha X^I + i\, S\, \gamma^- (\partial_\tau + \partial_\sigma) S] \tag{29}$$

where X^i denote the transverse coordinates of the string (i=1...8) X^I are internal coordinates (I=1,...16) which lie on a torus T^{16} whose radii are all equal to R, and $R^2 = \alpha'$.

The internal coordinates are constrained to be right-moving:

$$(\partial_\tau - \partial_\sigma) X^I = 0 \tag{30}$$

The spinor S^a is Majorana-Weyl and left-moving:

$$(\gamma^0 + \gamma^9)S = (1+\gamma^{11})S = 0$$
$$S^a = S^a(\tau+\sigma) \tag{31}$$

The effective action for a heterotic string moving in a background containing gravitons and anti-symmetric tensors may be deduced by using our previous techniques. We shall not do this, but simply mention the result: the effective action has general covariance in the transverse space and when written in terms of Ramond-Neveu-Schwarz fields processes a N = 1/2 supersymmetry on the world sheet.[14]

We shall now discuss in some detail the propagation in a background gauge field. This involves new features and rather novel manifestations of gauge invariance.

In the heterotic string, the massless vector modes are of two types. The first type corresponds to the diagonal gauge fields of $E_8 \times E_8$ or SO(32) and are constructed out of the following product states:

$$\tilde{a}^{\dagger}_{i1}\, a^{\dagger}_{I1}|0\rangle \tag{32}$$

Here $\tilde{a}^{\dagger}_{i1}$ denotes the standard creation operator in the left moving vector for the mode N = 1 in the i direction. $a^{\dagger}_{I1}$ is the creation operator in the right moving vector in an internal direction I. The vertex operator for such a massless vector particle coupling to the string is given by[15]

$$\frac{K}{\pi}\int_0^{\pi} d\sigma B^i_c(k,\tau-)\,\frac{\partial X^I}{\partial\tau+}\,e^{ikx}\zeta_{i1}(K) \tag{33}$$

where ζ is the polarization vector with I labelling the particular diagonal generator of $E_8 \times E_8$ or SO(32). The other quantities in (33) are defined as in eqs. (25).

The other massless modes are states of the form

$$\tilde{a}^{\dagger}_{i1}\quad |P^I\rangle \tag{34}$$

where $|P^I\rangle$ is the ground state of the right moving sector carrying an internal momentum P^I. P^I is a vector in the root space of $E_8 \times E_8$ or SO(32). Each such state corresponds to an off-diagonal component of the gauge field. The corresponding vertex operator is:[15]

$$\frac{K}{\pi}\int_0^\pi d\sigma B^i_C(k,\tau_-)e^{ik\,x}\colon e^{2iP^Ix^I}\colon C(P_I)\zeta_{i,P_I} \tag{35}$$

Here C(K) is an operator cocycle acting on right moving states with internal momenta P^I as:

$$C(K)|P\rangle = \varepsilon(K,P)|P\rangle \tag{36}$$

The two cocycle $\varepsilon(K,P)$ may be chosen to be bilinear and to satisfy

$$\begin{aligned}&\varepsilon(K,L)\varepsilon(L,K) = (-1)^{K\,L}\\ &\varepsilon(K,0) = -\varepsilon(K,-K) = 1\end{aligned} \tag{37}$$

We can now write down the interaction term of the string with background gauge fields. For diagonal fields one has

$$\sum_I\int d\tau d\sigma\left\{A_{iI}(x)(\partial_\tau-\partial_\sigma)x^i - i\partial_j A_{iI}(x)\bar{S}\gamma^{ij-}S\right\}(\partial_\tau+\partial_\sigma)x \tag{38}$$

while for off-diagonal fields $A_i(x,P_I)$ one has

$$\sum_{P^I}\int dJd\sigma\left\{A_i(x,P_I)(\partial_\tau-\partial_\sigma)x^i - i\,\partial_j A_i(x,P_I)\bar{S}\gamma^{ij-}S\right\}e^{iP_Ix^I}C(P_I)\,. \tag{39}$$

The gauge fields $A_{iI}(x)$ are analogs of Kaluza-Klein gauge fields in ordinary particle theories. The off-diagonal fields $A_i(x_1P_I)$ are, however, a special feature of strings.

The discussion of gauge invariance of the coupling of diagonal fields (given by eq. (38) is similar to that of ordinary Kaluza-Klein theories. Let us digress a bit and discuss the latter in a first quantized path integral approach.

Consider a particle moving on a five dimensional curved manifold described by a metric g_{ab} (a,b=1..5). The action is

$$S = \int d\tau g_{ab} \frac{dx^a}{d\tau} \frac{dx^b}{d\tau} \tag{40}$$

Suppose the fifth dimension $x^5=y$ curls up into a circle. Then the component of the metric $g_{a5}=A_a$ behaves as a U(1) gauge field. The amplitude $\langle x_2|x_1\rangle$ given by the path integral

$$\langle x_2|x_1\rangle = \int_{x(\tau)=x_1}^{x(\tau_2)=x_2} Dx(\tau) \exp(i\int d\tau g_{ab} \frac{dx^a}{d\tau} \frac{dx^b}{d\tau}) \tag{41}$$

is invariant under general coordinate transformations:

$$x^a \to x^{a'} = x^a + \varepsilon^a$$

$$g_{ab} \to g'_{ab} = \Delta_a \varepsilon_b + \Delta_b \varepsilon_a + g_{ab} \tag{42}$$

This is, of course, the statement that the second quantized field $\phi(x)$ describing the system is a scalar. Consider the special coordinate transformations;

$$\varepsilon^\mu = 0 \qquad \mu = 1,2,3,4$$

$$\varepsilon^5(x^a) = \varepsilon^5(x^\mu) \tag{43}$$

and the case where $g_{\mu\nu}=\eta_{\mu\nu}$. Then

$$g'_{\mu\nu} = g_{\mu\nu} \tag{44}$$

while

$$g'_{\mu 5} = g_{\mu 5} + \partial_\mu \alpha(x) \tag{45}$$

In the four dimensional world the relevant amplitude is $\langle x_2{}^\mu|x_1{}^\mu\rangle$ (μ=1,..4) since a charge of the compact dimension x^5 is not perceivable. Under the above transformation this amplitude transforms as:

$$\langle x_2{}^\mu|x_1{}^\mu\rangle \to \langle x_2{}^\mu|x_1{}^\mu\rangle \exp[i\int \partial_\mu \alpha(x) \frac{dx^\mu}{d\tau} \frac{dy}{d\tau}] \tag{46}$$

Consider the part of the amplitude obtained by summing over paths which have constant velocity in the fifth direction, i.e.

$$\frac{dy}{d\tau} = e = \text{constant} \tag{47}$$

Then (46) reads

$$\langle x_2 | x_1 \rangle \to e^{ie\alpha(x_2)} \langle x_2 | x_1 \rangle e^{-ie\alpha(x_1)} \tag{48}$$

which is just a U(1) gauge transformation with charge e. What is the meaning of this sub-amplitude in field theory language? The field $\phi(x^\mu,y)$ may be Fourier decomposed:

$$\phi(x^\mu,y) = \sum_q \phi_q(x^\mu)\, e^{iqy} \tag{49}$$

Under the general coordinate transformation of the special type (45), each ϕ_q transforms as

$$\phi_q(x^\mu) \to \phi_q{}^1(x^\mu) = \phi_q(x^\mu) e^{iq\alpha(x)} \tag{50}$$

since the whole field $\phi(x^\mu,y)$ is a scalar. Thus ϕ_q behaves as a charged field of charged q. Note that the field is not in general an eigenstate of change. Only if

$$\phi(x^\mu,y) \sim \phi(x^\mu)\, e^{iqy} \tag{51}$$

the four dimensional field $\phi(x^\mu)$ has a definite charge. However (51) is in a sense equivalent to the condition (47) in the path integral description, since the velocity in the y direction is given by

$$-\frac{i}{2}\, \phi^*(x,y)\overleftrightarrow{\partial}_y\phi(x,y)$$

and is a constant for $\phi(x,y)$ of the form given in eq. (51). The case of a dimensions with the d-4 dimensions forming a torus T^{d-4} is a trivial generalization of the above.

We are now ready to discuss the heterotic string. Consider a transformation

$$A_{iI}(x) \to A_{iI}(x) + \partial_i \alpha_I(x) \quad . \tag{52}$$

in the action (38). The change in the action is

$$\sum_I \int d\tau d\sigma \; \partial_i \alpha_I(x)(\partial_\tau - \partial_\sigma)X^i(\partial_\tau + \partial_\sigma)X^I \quad . \tag{53}$$

Consider the subamplitude for propagation of the string obtained by restricting paths by

$$\rho^I = (\partial_\tau + \partial_\sigma)X^I = \text{constant} \quad . \tag{54}$$

This is the string analog of Eq. (47). Eq. (52) now becomes

$$\sum_I \int d\sigma \; \alpha_I \cdot \rho^I \Big|_{\tau_i}^{\tau_2} - \int d\tau \; \alpha \; \rho^I \Big|_{\sigma=0}^{\sigma=\pi} \quad . \tag{55}$$

The second term in (55) vanishes by the periodicity condition for closed strings $X(0,\tau) = X(\pi,\tau)$. Thus the string amplitude transforms as

$$\Big\langle x_2(\sigma)|x_1(\sigma)\Big\rangle \to \exp\left(i \sum_I \int d\sigma \; \alpha_I(x(\sigma, \tau_1))\rho^I \right)$$

$$\Big\langle x_2(\sigma|x_1(\sigma)\Big\rangle \to \exp\left(-i \sum_I \int d\sigma \; \alpha_I(x(\sigma, \tau_1))\rho^I \right) \tag{56}$$

The string propagates as if constant charge densities ρ^I are distributed along the string.

Now consider transformations of the off-diagonal fields:

$$A_i(x,P_I) \to A_i(x,P_I) + \partial_i \alpha(x,P_I) \quad . \tag{57}$$

The string subamplitudes obtained by restricting to paths x^I = constant now transform as if there are uniform charge densities

$$\rho(P_I) = :e^{2iP_I x^I}:c(P_I) \tag{58}$$

distributed along the string.

The various charges obtained from the above charge densities in fact form the $E_8 \times E_8$ or SO(32) Lie algebra. Consider the charges[13]

$$K^I = \int d\sigma (\partial_\sigma + \partial_\tau) x^I$$

$$E(P_I) = \int \frac{dz}{2\pi i z} : e^{2iP_I x_I} : c(P_I) \qquad (59)$$

$$z = \exp(2i(\sigma+\tau))$$

where now the x^I's are not restricted in any fashion. These charges obey the commutation relations of the Lie algebra of $E_8 \times E_8$ (Chevelley basis).[13] K^I are the elements of the Cartan subalgebra, while $E(P_I)$ are the raising and lowering operators. This is in fact the reason why $A_{iI}(x)$ and $A_i(x, P_I)$ are the diagonal and off-diagonal gauge fields, and the string amplitudes transform covariantly under the linearized gauge transformations (52) and (57).

CONCLUSIONS

In this talk I have indicated how the interaction vertices of string theories ensure that a string would propagate in a non-trivial vacuum (containing condensates of massless field) according to linearized gauge and general coordinate invariance. The main task now is to understand how these invariances are present in their non-linear form. Hopefully some progress shall be made in the framework of covariant string field theories.

EPILOGUE

I shall end this talk with an interesting aspect of effective actions discussed here. As we have seen, the propagation of a single string in arbitrary background fields is described by non-linear sigma models on the world sheet. For the latter to be consistent, conformal invariance must be maintained at the quantum level. This means that all the beta-functions must vanish. It turns out that when one imposes these conditions (at least up to two-loop levels), massless background fields obey the classical equations of motion obtained from the zero slope limit field theory lagrangian.[14] In fact, for the heterotic string one obtains the well-known Chern-Simons terms as well. Thus, in particular, one has consistent string propagation only if the Einstein equations are obeyed by the metric, with the relevant energy momentum tensors of the other fields (Kalb-Ramond, Yang-Mills, etc.). This result may throw valuable light on the role of the zero-slope-limit field equations in the compactification of the string theory.

ACKNOWLEDGEMENTS

I would like to thank S.R. Wadia for collaboration and the members of the Theory Group at Tata Institute for Fundamental Research for hospitality (where most of the work was done). I also wish to thank the organizers and participants of this workshop for a very stimulating atmosphere.

REFERENCES

†Address after November 1: Department of Physics, California Institute of Technology, Pasadena, CA 91125

1. For a recent review, see J.H. Schwarz, Phys. Rep. 89C (1982) 223. Some earlier reviews are: J. Scherk, Rev. Mod. Phys. 47 (1975) 123; C. Rebbi, Phys. Rep. 12C (1974) 1.

2. S.R. Das and S.R. Wadia, Fermilab Preprint 85/90-T (May, 1985).
3. M. Kaku and J. Lykken, *Proceedings of Argonne Conference on Anomalies, Geometry and Topology* (World Scientific, Sept. 1985); T. Banks and M. Peskin, SLAC Preprint (July 1985); D. Friedan, University of Chicago Preprint (June 1985).
4. M. Kaku, talk at this workshop.
5. S. Raby, talk at this workshop.
6. M. Rubin, talk at this workshop; S.R. Das and M. Rubin (under preparation).
7. R.P. Feynman, Phys. Rev. 76 (1949) 769; Phys. Rev. 80 (1950) 440.
8. See discussion in C. Rebbi, Ref. 1.
9. T. Curtright and C. Zachos, Phys. Rev. Lett. 53 (1984) 1799.
10. S. Jain, R. Shankar and S.R. Wadia, TIFR Preprint TIFR-TH-85-3; D. Nemeschansky and S. Yankelowicz, Phys. Rev. Lett. 54 (1985) 620; E. Bergshoff, S. Ranjdbar-Daemi, A. Salam, H. Sarmadi and E. Sezgin, ICTP Preprint (May 1985).
11. J.H. Schwarz, Ref. 1.
12. The following action has been derived directly in A. Sen, Fermilab Preprint 85/60-T.
13. D.J. Gross, J. Harvey, E. Martinec and R. Rohm, Phys. Rev. Lett. 54 (1985) 502; Nucl. Phys. B256 (1985) 253.
14. A. Sen, Fermilab Preprint 85/77-T; C.A. Callan, E.J. Martinec, M. Perry and D. Friedan, Princeton preprint.
15. J. Harvey, *Proceedings of Argonne Conference on Anomalies, Geometry and Topology* (World Scientific, Sept. 1985).

Symmetry breaking on compact manifolds

J. D. Breit
University of Pennsylvania
Philadelphia, Pennsylvania 19104

The compactification of superstring theories on Calabi-Yau manifolds by Candelas, Horowitz, Strominger, and Witten is reviewed along with some problems with supersymmetric grand unified theories with gauge group E_6. A method of breaking the E_6 symmetry on non-simply connected manifolds is presented that can solve some of these problems.

In this talk I'd like to tell you about some work done with Burt Ovrut and Gino Segre[1] on a new method[2,3] of breaking gauge symmetries that may be useful in string theories. This method exploits the fact that when we compactify a higher-dimensional theory the compact manifold is not necessarily simply connected, and hence the vacuum configuration for the gauge fields does not have to be trivial. The gauge symmetry can then be broken in such a way that a fine tuning problem due to proton decay can be avoided.[1,3,4] Since most of the talks here have been about ten-dimensional superstring theories per se rather than their connection with the four-dimensional field theories we see at low energies, I'd like to first review for you compactification, in particular the Calabi-Yau manifolds studied by Candelas, Horowitz, Strominger, and Witten[2] (CHSW). I won't go into great detail, my aim being to make plausible some of their results we shall need for the symmetry breaking.

Compactification on Calabi-Yau manifolds. The starting point for CHSW is ten-dimensional N=1 supergravity modified by appropriate chern-simons forms. The fields are the graviton and gravitino, dilaton and its majorana spinor partner, and an antisymmetric tensor and a (1,1/2) multiplet of gauge fields in the adjoint representation of either $E_8 \times E_8$ or SO_{32}. We won't need to go into the details of their analysis except to note that if we write the antisymmetric tensor as a two form B, its field strength $H = dB - \omega_{3Y} + \omega_{3L}$, where $d\omega_{3Y} = tr(F \wedge F)$, $d\,\omega_{3L} = tr(R \wedge R)$, and tr is the trace in the vector representation. They assume that ten-dimensional space compactifies to $M \times K$, where M is a maximally symmetric four-dimensional space and K is a compact six dimensional manifold. They further demand that the four-dimensional theory have an unbroken N=1 supersymmetry. These conditions place restrictions on the background field configurations after compactification since the vacuum must be invariant under the supersymmetry transformations.

By studying the effects of a supersymmetry transformation, CHSW found various integrability conditions, in particular, the maximally symmetric four-dimensional manifold M must be flat Minkowski space and the three form H must satisfy a nonlinear integral equation. This equation has the solution H=0, which they chose. The consequences of H=0 are quite restrictive. It implies the existence of a covariantly constant spinor on K which then implies that K must be Ricci flat and Kähler (the line element $ds^2 = g(z,\bar{z})dz \wedge d\bar{z}$ is a closed two form known as the Kähler form). Since $dH = ddB - d\omega_{3Y} + d\omega_{3L} = 0$, we have $tr(F \wedge F) = tr\,(R \wedge R)$.

If a spinor is covariantly constant, when we parallel transport it around a closed loop it ends up pointed the same way as it started. On general manifolds this does not happen. Consider for example the two sphere. Take a vector at the north pole pointed along the Greenwich meridien and parallel transport it to the south pole along that meridien. Now parallel transport it along the 90th meridien back to the north pole. The vector is now pointed in the opposite direction. The rotations induced by parallel transporting along different paths form a group known as the holonomy group. For the two sphere the holonomy group is O_2; for the plane it is just the identity element. For an n-dimensional manifold one expects in general that the holonomy will be O_n. If there is a covariantly constant spinor or vector the holonomy group will be some subgroup of O_n. In our case, since $O_6 = SU_4$ and the covariantly constant spinor picks out a direction that is not rotated by parallel transport, the holonomy group is the little group of SU_4, SU_3. So K must be a Ricci flat manifold with SU_3 holonomy; such manifolds are known as Calabi-Yau spaces.

As Paul Frampton mentioned in his talk, in order to have chiral fermions in four dimensions we must have some background gauge fields in the compactified dimensions. The condition that $tr(F\wedge F) = tr(R\wedge R)$ specifies the form of those backgound gauge fields, namely we set the background gauge field equal to the spin connection by embedding the SU_3 spin connection in either $E_8 \times E_8$ or SO_{32}. The remaining gauge symmetry is the subgoup that commutes with SU_3. This embedding breaks SO_{32} down to SO_{26}, which leads to only vector-like representations. For $E_8 \times E_8$, on the other hand, the gauge symmetry is broken down to $E_8 \times E_6$, and the 27 dimensional representation of E_6 leads to chiral fermions. The gauge fields are in the 78 of E_6 and the unbroken E_8 is a hidden sector.

Finally we have to see how many generations of fermions remain after compactification. Suppose a field σ satisfies the equation $\nabla^2 \sigma = 0$ in ten dimensions, or $\nabla^2_4 \sigma + \nabla^2_6 \sigma = 0$ on $M \times K$. Expand in eigenfunctions on K, $\nabla^2_6 \sigma = \lambda \sigma$; then on M we have $(\nabla^2 + \lambda)\, \sigma = 0$ and λ acts as a mass. Since the compact manifolds are supposed to be quite small, λ is a very large mass. Therefore the only light particles on M are those for which $\lambda = 0$. In other words we can count light particles on M by counting harmonic forms on K (a form f is harmonic if it satisfies a generalized Laplace's equation; on a compact manifold $df = d{*}f = 0$, that is, f is both closed and co-closed).

We can count harmonic forms by using cohomology theory. Suppose we have a closed form f, that is df=0. Locally we can always write f as an exact form, f = dg, but we can't always define g globally. So the closed forms that cannot be written as an exact form give us information about the global topology. Define an equivalence class: two closed forms are equivalent if they differ by an exact form. The dimension of this class is called the Betti number b_n, and b_n gives the number of harmonic n forms. Let's look at the two sphere for an example. No zero form can be exact since if f is a zero form and f = dg, g would have to be a −1 form. So all we have to do is find the closed 0 forms. These are just the functions on the sphere whose derivatives vanish, the constant functions. The constant functions can be equal to any real number, so b_0 is just the dimension of the real number line or 1. Suppose we have a closed one form f. Integrate df over a portion of the sphere; the integral vanishes since df does. By Stokes theorem this integral is equal to the integral of f over the boundary of the patch. Taking all possible portions of the sphere, we have that $\int f$ over all closed paths vanishes and hence $f = dg$. So all closed one forms are exact and $b_1 = 0$. Finally, all two forms are closed since there are no three forms on the two sphere. If $f = dg$ then $\int f$ over the sphere equals $\int g$ over the boundary of the sphere which vanishes since a sphere has no boundary. So if we find a two form f for which $\int f \neq 0$ then f is closed but not exact. Such a form is given by the volume form, and so $b_2 = 1$. Note that the Euler characteristic of a manifold is defined as the alternating sum of the Betti numbers, so in this case it is 2.

Suppose we compactify on $M \times S^2$. A scalar field on $M \times S^2$ will lead to a massless scalar field on M since $b_0 = 1$. An antisymmetric tensor field could lead to a scalar field on M (neither index in M), a vector field (one index in M), and an antisymmetric tensor (both indices in M). The scalar field and antisymmetric tensor field are massless since $b_2 = 1$ and $b_0 = 1$, but the vector field is massive since $b_1 = 0$. On Calabi-Yau manifolds things are more complicated. The manifolds are complex and instead of n-forms we have (n, m) forms, where n counts the number of dz's and m the $d\bar{z}$'s. Instead of Betti numbers we have hodge numbers $b_{n,m}$ which give the number of harmonic (n, m) forms. The basic idea is similar, however, and CHSW are able to specify all the massless fields that remain after compactification. In particular they find that the number of right-handed fermions is given by $b_{1,1}$ and the left-handed by $b_{1,2}$. The difference between them is just half the Euler characteristic χ of the manifold.

Since all the Calabi-Yau manifolds they studied had very large Euler characteristics, the number of fermion generations comes out much too high. They also found, however, that these manifolds had discrete symmetries, and identifying points related by the discrete symmetry reduces χ. For example, one manifold given by a quintic equation in CP^4 gave $\chi = -200$, but also has a $Z_5 \times Z_5$ discrete symmetry. The manifold $K/Z_5 \times Z_5$ has $\chi = -200/25 = -8$ and hence an acceptable four generations. They also noted that modding out the discrete symmetry leads to a multiply connected manifold, and on such manifolds a new way of breaking gauge symmetries is possible.

E_6 *Phenomenology.* The group E_6 once enjoyed a vogue in grand unified models and is fairly well studied.[5] In our case there are two important differences with earlier work: the theory is supersymmetric and we have only 27-dimensional representations for the Higgs bosons. Supersymmetry implies the existence of color triplet Higgs bosons that mediate fast proton decay through QQH couplings. The absence of higher-dimensional representations means that we cannot break the symmetry and give fermions masses by using Higgs bosons in the 78, 351, and 351' representations as is usually done.

To simplify the discussion, let's write the members of the 27 in a suggestive way by considering the decomposition of E_6 into $SO_{10} \times U_1$ and then SO_{10} into $SU_5 \times U_1$. We label states by the notation $[SO_{10}, SU_5]$. For the 27 we have

$$27 = [1,1] + [10,5] + [10,\bar{5}] + [16,1] + [16,\bar{5}] + [16,10]$$

and let

$$[1,1] = N_0 \qquad [16,1] = \nu^c$$

$$[10,5] = \begin{pmatrix} N^c \\ E^c \\ D \end{pmatrix} \qquad [10,\bar{5}] = \begin{pmatrix} N \\ E \\ D^c \end{pmatrix} \qquad [16,\bar{5}] = \begin{pmatrix} \nu \\ e \\ d^c \end{pmatrix}$$

$$[16,10] = \begin{pmatrix} & u & \\ e^c & & u^c \\ & d & \end{pmatrix}$$

The usual fermions are contained in the $[16,10]$ and $[16,\bar{5}]$. We also have two extra neutrinos, the N_0 and ν^c, and the extra isodoublet and color triplet N, E, and D, and all of these particles have scalar superpartners. In E_6 we have the allowed coupling $27 \times 27 \times 27$; in terms of the above classification this coupling becomes $[1,1] \times [10,5] \times [10,\bar{5}]$, $[16,1] \times [10,5] \times [16,\bar{5}]$, $[16,10] \times [16,\bar{5}] \times [10,\bar{5}]$, $[16,10] \times [16,10] \times [10,5]$.

Looking at these couplings, we can see an immediate problem. Terms like udD and $u^c d^c D^c$ can lead to instantaneous proton decay through the exchange of a color triplet squark. The obvious solution is to make the D and D^c very heavy. This solution, however, leads to another problem. The fermions in the [16,10] and [16,$\overline{5}$] get their masses through couplings to the [10,5] and [10,$\overline{5}$], in particular to the N, N^c, E, and E^c. So we expect these sleptons to have vev's on the order of the electroweak scale. It is very difficult to see how these particles can have such small vev's while the D squark is superheavy without a great deal of fine tuning. The new method of symmetry breaking gives a possible solution to this problem.

Symmetry breaking. As we have seen, the simply connected Calabi-Yau manifolds studied by CHSW give rise to too many fermion generations. To get an acceptable number of generations, they formed quotient spaces K/D, where D is a freely acting discrete group. The manifold is now multiply connected, and they proposed using this property to break the E_6 symmetry in the following way. Consider a Wilson loop on K

$$U = P\exp\{ i \oint A \cdot dl \}$$

The Wilson loops U are elements of the adjoint representation (78) of E_6. If $U \neq 1$ the symmetry is broken to that subgroup G of E_6 which commutes with U. Now on a simply connected manifold we can always contract a loop to a point, so

$$\oint A \cdot dl = \int F \cdot dS$$

In the vacuum F = 0 and hence U = 1. On a multiply connected manifold loops cannot always be shrunk to a point, and so we are not guaranteed to find U = 1. In other words, on multiply connected manifolds there can exist nontrivial background gauge field configurations that give F = 0 and hence are degenerate with the usual vacuum (A = 0). These vacua break the gauge symmetry. Determining which vacuum is actually favored requires a calculation of quantum corrections and is a very difficult problem. What we can do without too much effort is find constraints on the allowed symmetry-breaking patterns. Rather than describe the method in general, I'd like to illustrate the salient points with a simple example.

Consider SU_3 gauge fields on the compact manifold $RP^2 = S^2/Z^2$, the real projective plane. Let's first see that there exist closed loops on RP^2 that cannot be contracted to a point. Take a curve on the sphere that begins at the north pole and ends at the south pole. Although this is not a closed curve on S^2, on S^2/Z^2 we identify antipodal points, and hence the north and south poles, so it is a closed path. This curve cannot be contracted to a point because on S^2 its ends are tied to antipodal points, which can never be brought together. Suppose we went around this path twice on RP^2 ; that is, we start at the north pole, go to the south pole, reappear at the north pole, and go to the south pole again. On the sphere this path corresponds to, say, following the Greenwich meridien down to the south pole and coming back up the international date line, a contractible loop. So we have noncontractible loops on RP^2 , but if we go around them twice they can be contracted to a point. Similarly for a multiply connected manifold K/Z^n, if we go around a loop n times it can be contracted.

What does all this imply for the Wilson loops U? Since Z^2 is abelian, they can be written as diagonal 3x3 matrices with unit determinant. Since loops are contractible if we go around them twice, $U^2 = 1$. The possibilities are:

$$(a)\begin{pmatrix}1&&\\&1&\\&&1\end{pmatrix}\quad(b)\begin{pmatrix}1&&\\&-1&\\&&-1\end{pmatrix}\quad(c)\begin{pmatrix}-1&&\\&1&\\&&-1\end{pmatrix}\quad(d)\begin{pmatrix}-1&&\\&-1&\\&&1\end{pmatrix}$$

We can now simply read off the symmetry breaking. In case (a) it is still SU_3 and for (b)-(d) it is $SU_2{\times}U_1$. Note that for abelian discrete groups the rank of the group is preserved.

For most of the Calabi-Yau manifolds given in CHSW the only (1,1) form was the Kähler form, and hence there was only one right-handed 27. An explicit $27{\times}\overline{27}$ mass term is forbidden since these fields are all zero modes, but we can use the Wilson loops as an effective 78 to generate a mass term $27 \times 78 \times \overline{27}$. One of the 27's can pair off with the $\overline{27}$ and become heavy. In our example one of the 3's pairs off with a $\overline{3}$ via the 8 given by the Wilson loop. To see which elements of the 3 become heavy, we just look at the U's above for entries that are different from one. For case (a) none of the 3 becomes heavy, while for (b)-(d) only one remains massless.

One can apply the same techniques to E_6, but 27x27 matrices become somewhat awkward. We prefer to use a different method. The representations of groups can be given in terms of their roots and weights.[6] One can analyze rank-preserving symmetry breaking by looking at the inner products of a symmetry breaking vector $c_i H^i$ with the roots, where H^i are the generators of the Cartan subalgebra. Those roots which have a nonvanishing inner product with this vector correspond to gauge bosons that become massive. For SU_3 the nonzero roots are (2 −1), (1 1), and (−1 2). Write the symmetry-breaking vector as (a b). Since $U^2 = 1$, a and b must be elements of the integers under addition mod 2, that is 0 or 1. So the allowed symmetry breaking directions are (0 0), (1 0), (0 1), or (1 1). (0 0) corresponds to case (a) above, and the rest to (b)-(d). For instance the inner products ((1 1),(2 −1)) = 1, ((1 1),(1 1)) = 2 = 0 mod 2, and ((1 1),(−1 2)) = 1; so the roots (2 −1) and (−1 2) correspond to massive vector mesons and the symmetry is broken to $SU_2 \times U_1$. To find which elements of the 3 that pairs off with the $\overline{3}$ acquire a mass, we look at the inner product of the symmetry breaking direction with the weights of the 3 of SU_3, (1 0), (−1 1), and (0 −1). For example, ((1 1),(1 0)) = 1, ((1 1),(−1 1)) = 0, and ((1 1),(0 −1))=1; so two of the 3 become massive.

One can readily extend this procedure to E_6. This group has rank six, so the symmetry breaking direction has the form (a b c d e f). We want to preserve an unbroken $SU_3 \times SU_2 \times U_1$, so the inner products of (a b c d e f) with the color and weak isospin roots must vanish. This constraint implies that the symmetry-breaking direction is of the form (−c c a b c 0). It is then straightforward to use the tables given in Slansky's Physics Report[6] to find the dependence on a, b, and c of the gauge boson masses and the masses of the 27 and $\overline{27}$ that pair off. Since a, b, and c are constrained by the fact that for a given discrete group $U^n = 1$, one can enumerate all the allowed symmetry breaking patterns.

In particular, for the manifold found by CHSW that gave four generations, we find that one can break E_6 to $SU_3 \times SU_2 \times SU_2 \times U_1 \times U_1$ in such a way that only two isodoublets and one singlet remain light in the 27 that pairs off with the $\overline{27}$. This result potentially solves the fine-tuning problem.[1,3,4] If all the Higgs come from this 27, then since all its color triplets have become heavy there is no instantaneous proton decay. If we call this 27 H and the others Ψ, we must forbid couplings of the form $\Psi\Psi\Psi$, since the color triplet squarks in Ψ can still lead to proton decay. We can forbid such couplings by positing an extra discrete symmetry, say $\Psi \to -\Psi$. On many Calabi-Yau manifolds there are such additional discrete symmetries.[3] None as yet is known to forbid the "bad" Yukawa couplings, but there is hope that some will be found.

Acknowledgment. It is a pleasure to thank Gino Segre` for many helpful discussions about topology and differential forms.

References

1. J. D. Breit, Burt A. Ovrut, and Gino C. Segre`, Physics Letters 158B, 33 (1985).
2. P. Candelas, G. Horowitz, A. Strominger, and E. Witten, "Vacuum configurations for superstrings," Nucl. Phys. B (to be published).
3. E. Witten, "Symmetry breaking patterns in superstring models," Nucl. Phys. B (to be published).
4. A. Sen, Phys. Rev. Lett. 55,33 (1985).
5. F. Gursey, P. Ramond, and P. Sikivie, Phys. Lett. 60B, 177 (1976); F. Gursey and M. Serdaroglu, Nuovo Cimento 65A, 337 (1981).
6. R. Slansky, Phys. Rep. 79, 1 (1981).

Finiteness of One Loop Parity Conserving Superstring Amplitudes

L. Clavelli

Department of Physics and Astronomy
University of Alabama
University, AL 35486*

and

Department of Physics
University of Indiana
Bloomington, IN 47405

Abstract

We prove, in the Lorentz covariant formulation of the type I Superstring theory, that the parity conserving one loop amplitudes with arbitrary numbers of external gauge bosons are finite if the internal symmetry group is SO(32).

In the past year a great amount of interest has been rekindled in supersymmetric string theories by the prospect of developing an infinity free and anomaly free theory of fundamental interactions including Yang Mills and gravitational forces. Of particular interest from some points of view is the SO(32) type 1 theory which includes both open and closed strings. Using a non-

*Present address

manifestly Lorentz invariant light cone formulation, Green and Schwarz[2] have recently shown that the one loop four point function is finite. One might expect from this, using a heuristic duality argument, that all the higher N point functions are finite at the one loop level. However in the light cone formulation, the finiteness of the four point function appears to depend explicitly on the number of external particles.

In this article we show directly the finiteness of all the parity conserving one loop N point functions. The vanishing of the one loop 2 and 3 point functions follows in the Lorentz covariant formulation from simple relations among Jacobi theta functions.

The type 1 super string theory corresponds to the Ramond-Neveu-Schwarz model projected onto the even G parity bosonic sector and the Majorana-Weyl fermion sector[3].

Up to internal symmetry factors the one loop N point functions in D dimensions with external gauge bosons are given by expressions of the form

$$L = g^N d^Dp \, \mathrm{Tr} \, P \prod_{i=1}^{N} [V(k_i,1)\Delta] \tag{1}$$

With bosons circulating in the loop the vertex for gauge boson emission is

$$V(k_i,\rho_i) = V_B(k_i,\rho_i)$$

$$= e^{ik_i\cdot Q(\rho_i)} [\zeta_i\cdot P(\rho_i) + k_i\cdot H(\rho_i)\zeta_i\cdot H(\rho_i)] \quad (2)$$

and the string propagator is

$$\Delta = \Delta_B = \frac{1}{L_0-1}\,(1+\Omega)\left(\frac{1+G}{2}\right) = \int_0^1 \frac{dx}{x}\, x^{L_0-1}\,(1+\Omega)\left(\frac{1+G}{2}\right) \quad (3)$$

Here Ω is the twist operator given (up to a gauge transformation) for an orthogonal group internal symmetry by

$$\Omega = -e^{i\pi(L_0-p_0^2/2)} \quad (4)$$

and G is the G parity operator

$$G = e^{2\pi i(L_0-p_0^2/2)} \quad (5)$$

P, in eq. 1 is the Brink Olive projection operator onto positive norm states. The amplitude is given by a D dimensional loop momentum integration times a trace over excited states of the circulating string. With fermions circulating, the amplitude takes the same form (except for a fermionic loop factor of -1) with the substitutions

$$V(k_i,\rho_i) = V_F(k_i,\rho_i) = e^{ik_i\cdot Q(\rho_i)}\, \zeta_i\cdot\Gamma(\rho_i)/(\sqrt{2}i) \quad (6)$$

$$\Delta = \Delta_F = \frac{1}{F_0}(1+\Omega)\ \frac{(1+\Gamma^{11})}{2} \tag{7}$$

In the following we will not treat the parity violating term Γ^{11} in the propagator. Possible divergences due to these terms cannot cancel divergences in the parity conserving terms. The parity conserving loop is also given in an alternate (F_1) formulation by

$$V_F(k_i,\rho) = e^{ik_i\cdot Q(\rho)}[\zeta_i\cdot P(\rho) - \frac{1}{2}\, k_i\cdot\Gamma(\rho)\zeta_i\cdot\Gamma(\rho)] \tag{8}$$

$$\Delta_F = \frac{1}{L_0}(1+\Omega) \tag{9}$$

providing one supplies an extra factor of $\frac{1}{4}$ on the right hand side of eq. 1. With fermionic states circulating, the trace in eq. 1 is, of course, taken also over the Dirac matrices.

Loops with a positive even number of twists in eq. 1 are finite except on the poles of intermediate graviton states and need not be considered here. Loops with an odd number of twists are equivalent to the Mobius loop with a single twist if the integration range of the propagator variables is appropriately adjusted as discussed below.

It is convenient to replace the loop momentum

integral in eq. 1 by a trace over canonical zeroth mode oscillators using the limiting procedure of ref 4. Then the bosonic loop takes the form

$$L_B = g^N \int_0^1 d\Omega \; \mathrm{Tr}\; w^{L_0-1} \; P(1+\Omega)\left(\frac{1+G}{2}\right)\prod_{i=1}^{N} V_B(k_i,\rho_i,a,a^\dagger) \quad (10)$$

where

$$d\Omega = \frac{dw}{w} \prod_{i=2}^{N} \frac{d\rho_i}{\rho_i} \quad (11)$$

The arguments $a,a^\dagger$ in eq. 10 refer collectively to the boson oscillators in Q_μ and P_μ and to the fermion oscillators in H_μ. The string coordinate $Q_\mu(\rho,a,a^\dagger)$ and conjugate momentum $P_\mu(\rho,a,a^\dagger)$ are

$$Q_\mu(\rho,a,a^+) = \sum_{n=0}^{\infty}\left(\frac{\Gamma(n+2\epsilon)}{n!}\right)^{1/2}(a_\mu^n \rho^{-n-\epsilon} + a_\mu^{n\dagger}\rho^{n+\epsilon}) \quad (12)$$

$$P_\mu(\rho,a,a^+) = i\,\rho\,\frac{d}{d\rho}\,Q_\mu(\rho,a,a^\dagger) \quad (13)$$

After evaluating the trace in eq. 10 one takes the limit $\epsilon \to 0$ using the relation

$$\lim_{\epsilon\to 0} e^{-(\Sigma_i k_i)^2/4\epsilon} \cdot (4\pi\epsilon)^{-D/2} = \delta^D(\Sigma k_i) \quad (14)$$

Using the results of ref 4 appendix C, the four terms in eq. 10 can be written

$$L_B = \frac{2^N}{2} d\Omega \{ (\mathrm{Tr}\ \lambda^\circ)[F_{NS}(w) T_{NS}(\rho_i, w) + F_{NS}(we^{2\pi i}) T_{NS}(\rho_i, we^{2\pi i})]$$

$$- F_{NS}(-w)\ T_{NS}(\rho_i, -w) - F_{NS}(-we^{2\pi i}) T_{NS}(\rho_i, -we^{2\pi i}) \} \qquad (15)$$

The projected Neveu Schwarz partition function is

$$F_{NS}(w) = \frac{w^{-1/2}}{w} \prod_{n=1}^{\infty} \left(\frac{1+w^{n-1/2}}{1-w^n} \right)^{D-2} \qquad (16)$$

We have taken the liberty of including in this a factor of 1/w which is more properly a part of $d\Omega$. This gives a differential $d\Omega$ which is then the same for boson and fermion loops. We assume the effect of the physical state projection operator is such as to maintain the form of eq 16 in the Mobius as well as in the planar loop. We have also defined

$$T_{NS}(\rho_i, w) = (-\epsilon\ \ln|w|)^{-D} \langle 0| \prod_{i=1}^{N} V_B(\rho_i, d(w), \bar{d}(w)) |0\rangle \qquad (17)$$

with

$$d^n(w) = \frac{a^n}{1 \pm w^{n-J}} + a'^{n\dagger} \tag{18}$$

$$\bar{d}^n(w) = a^{n\dagger} + \frac{a'^n w^{n-J}}{1 \pm w^{n-J}} \tag{19}$$

The + and - signs referring respectively to the fermionic and bosonic oscillators which have respectively $J = -\frac{1}{2}$ and $J = -\epsilon$. The zeroth modes in eq. 17 are understood to depend only on the absolute values of ρ_i and w.

Similarly the parity conserving fermion loop with N external gauge bosons is

$$L_F = -\frac{g^N}{2} d\Omega \ \{(\mathrm{Tr}\ \lambda^\circ) F_R(w) T_R(\rho_i, w) - F_R(-w) T_R(\rho_i, -w)\} \tag{20}$$

where the projected partition function in the Ramond sector is

$$F_R(w) = 2^{\frac{D-2}{2}} \prod_{n=1}^{\infty} \left(\frac{1+w^n}{1-w^n}\right)^{D-2} \tag{21}$$

and

$$T_R(\rho_i, w) = (-\epsilon\ \ell n|w|)^{-D} \langle 0| \prod_{i=1}^{N} V_F(\rho_i, d(w), \bar{d}(w)) |0\rangle \tag{22}$$

The vertices V_F of eq. 8 appear in eq. 22 with their fermionic and bosonic oscillators transformed by eqs. 18 and 19 with, effectively, J = 0 and $-\epsilon$ respectively. The vacuum expectation value in 22 is defined to include a normalized Dirac trace.

$$\langle 0| f(\gamma_\mu)|0\rangle \equiv 2^{-D/2} \text{ tr } f(\gamma_\mu)$$

The vacuum expectation values in eq. 17 and 22 are completely determined by the elementary correlations of pairs of conformal fields. The automorphisms of Jacobi theta functions relate these to similar correlations with transformed variables

$$\rho_i' = e^{2\pi i\, \ln\rho_i/\ln w} \tag{23}$$

$$w' = e^{4\pi^2/\ln w} \tag{24}$$

$$\rho_i'' = (\rho_i')^{1/2} \tag{25}$$

$$w'' = (w')^{1/4} \tag{26}$$

More specifically, with

$$\tau = \frac{\ln w}{2\pi i}, \tag{27}$$

one finds

$$<0|Q_\mu(\rho_i,d,\bar{d})Q_\nu(\rho_j,d,\bar{d})|0>$$

$$=<0|\hat{Q}_\mu(\rho_i{}',d',\bar{d}')\hat{Q}_\nu(\rho_j{}',d',\bar{d}')|0>$$

$$+ g_{\mu\nu}\{-\frac{1}{\epsilon^2 \ell n w} - \frac{1}{2\epsilon} - \frac{\pi i\tau}{6} - \ell n\tau\} \qquad (28a)$$

$$<0|P_\mu(\rho_i,d,\bar{d})Q_\nu(\rho_j,d,\bar{d})|0>$$

$$= \tau^{-1}<0|\hat{P}_\mu(\rho_i{}',d',\bar{d}')\hat{Q}_\nu(\rho_j{}',d',\bar{d}') \qquad (28b)$$

$$<0|P_\mu(\rho_i,d,\bar{d})P_\nu(\rho_j,d,\bar{d})|0>$$

$$= \tau^{-2}<0|\hat{P}_\mu(\rho_i{}',d',\bar{d}')\hat{P}_\nu(\rho_j{}',d',\bar{d}')|0> \qquad (28c)$$

Eqs. 28b and 28c follow by a simple differentiation of eq. 28a. Here we have defined

$$\hat{Q}_\mu(x,d,\bar{d}) = \nabla(2\epsilon)^{1/2}\ (a^\circ_\mu\ x^{-\epsilon} + a^{\circ\dagger}_\mu\ x^\epsilon) + \sum_{n=1}^{\infty}(\frac{\nabla(n+2\epsilon)}{n!})^{1/2}$$

$$\cdot(d^n x^{-n-\epsilon} + \bar{d}^n\ x^{n+\epsilon}) \qquad (29)$$

$$\hat{P}_\mu(x,d,\bar{d}) = ix\frac{d}{dx}\hat{Q}_\mu(x,d,\bar{d}) \qquad (30)$$

i.e., in $\hat{Q}_\mu$, by definition, the zeroth mode remains

untransformed by eqs. 18 and 19. The prime on d and $\bar{d}$ indicates that the argument w in eqs 18 and 19 is replaced by w'

$$d' \equiv d(w') \ , \quad \bar{d}' \equiv \bar{d}(w') \tag{31}$$

The Neveu Schwarz field H_μ satisfies

$$\langle 0| H_\mu(\rho_i,d,\bar{d}) H_\nu(\rho_j,d,\bar{d})|0\rangle$$

$$= \tau^{-1}\langle 0| H_\mu(\rho_i', d',\bar{d}') H_\nu(\rho_j',d',\bar{d}')|0\rangle \tag{32}$$

The non singular, ρ independent terms in the curly bracket of eq. 28a do not contribute to the loop amplitude with massless external particles since the $Q_\mu Q_v$ correlations only occur in the combination

$$\sum_{i<j} k_{i_\mu} k_{j_\nu} \langle 0| Q_\mu(\rho_i,d,\bar{d}) \ Q_\nu(\rho_j,d,\bar{d})|0\rangle \tag{33}$$

and the total four momentum is conserved. The singular terms in the curly bracket only affect the coefficient of the momentum conserving δ function.

Equations 28 and 32 imply that the vacuum expectation value of eq. 17 is simply related to another vacuum expectation value in which each $Q_\mu(\rho,d,\bar{d})$ is replaced by $\hat{Q}_\mu(\rho_i', d', \bar{d}')$, each $P_\mu(\rho_i, d, \bar{d})$ is replaced by $\tau^{-1}\hat{P}_\mu(\rho_i',d',\bar{d}')$ and each $H_\mu(\rho_i,d,\bar{d})$ is replaced by $\tau^{-1/2} H_\mu(\rho_i', d',\bar{d}')$.

More specifically, we define parallel to eq. 17

$$\hat{T}_{NS}(\rho_i',w') \equiv (4\pi\epsilon)^{-D/2} \langle 0| \prod_{i=1}^{N} \hat{V}_B(\rho_i',d',\bar{d}')|0\rangle \tag{33a}$$

$$\hat{T}_R(\rho_i',w') \equiv (4\pi\epsilon)^{-D/2} \langle 0| \prod_{i=1}^{N} \hat{V}_F(\rho_i',d',\bar{d}')|0\rangle \tag{33b}$$

where each Q_μ, P_μ in V_B or V_F has been replaced by $\hat{Q}_\mu$, $\hat{P}_\mu$, to form $\hat{V}_B$ or $\hat{V}_F$. Then

$$T_{NS}(\rho_i,w) = (\tfrac{\tau}{i})^{-D/2}\, \tau^{-N}\, \hat{T}_{NS}(\rho_i',w') \tag{34}$$

Under the same transformation the Neveu Schwarz partition function satisfies

$$F_{NS}(w) = (\tfrac{\tau}{i})^{D/2-1} F_{NS}(w') \tag{35}$$

and the differential transforms as

$$d\Omega = \tau^{N+1}\, d\Omega' \tag{36}$$

where $d\Omega'$ is given by eq. 11 in terms of primed variables. Putting eqs. 34, 35, and 36 together implies for the first term in eq. 15

$$d\Omega\ F_{NS}(w)\ T_{NS}(\rho_i,w) = i\ d\Omega'\ F_{NS}(w')\hat{T}_{NS}(\rho_i',w') \tag{37}$$

One can write similar relations for the correlations occuring in the Mobius loop and G parity projected loops. The results are that, in the Mobius loop, the Neveu Schwarz field correlation transforms into the G parity projected NS correlation and vice versa. Similarly the Planar Ramond Correlation transforms into the G parity projected planar Neveu Schwarz correlation. i.e. if

$$<0|H_\mu(\rho_i,d,\bar{d})H_\nu(\rho_j,d,\bar{d})|0> = g_{\mu\nu}\ \chi(\rho_j/\rho_i,w) \tag{38}$$

and

$$<0|\ \frac{\Gamma_\mu(\rho_i,d,\bar{d})}{i\sqrt{2}}\ \frac{\Gamma_\nu(\rho_j,d,\bar{d})}{i\sqrt{2}}|0> = g_{\mu\nu}\ \chi_0(\rho_j/\rho_i,w) \tag{39}$$

then

$$\chi(x,w) = \tau^{-1}\ \chi(x',w') \tag{40a}$$

$$\chi(x,we^{2\pi i}) = \tau^{-1}\ \chi_0(x',w') \tag{40b}$$

$$\chi_0(x,w) = \tau^{-1}\ \chi(x',w'e^{2\pi i}) \tag{40c}$$

$$\chi(x,-w) = (2\tau)^{-1}\ \chi(x'',-w''e^{2\pi i}) \tag{40d}$$

$$\chi(x,-we^{2\pi i}) = (2\tau)^{-1}\ \chi(x'',-w'') \tag{40e}$$

$$\chi_0(x,-w) = (2\tau)^{-1}\ \chi_0(x'',-w'') \tag{40f}$$

The Mobius loop involves correlations of fields with negative w

$$\langle 0|Q_\mu(\rho_i,d(-w),\bar{d}(-w)),Q_\nu(\rho_j,d(-w),\bar{d}(-w))|0\rangle$$

$$= \langle 0|\hat{Q}_\mu(\rho_i{}'',d(-w''),\bar{d}(-w''))\hat{Q}_\nu(\rho_j{}'',d(-w''),\ \bar{d}(-w''))|0\rangle$$

$$+g_{\mu\nu}\{-\frac{1}{\epsilon^2 \ln w} - \frac{1}{2\epsilon} - \frac{\pi i\tau}{6} - \ln 2\tau\} \tag{41}$$

Since the twist operator of eq. 4 involves no zeroth modes, the d's of negative argument in eq. 41 and elsewhere are defined so that the zeroth mode depends only on the absolute value of the argument. By differentiating eq. 41 with respect to $\ln \rho_i$ and/or $\ln \rho_j$ we find analogs of eqs. 28b and c with primes replaced by double primes on the right hand side and τ replaced by 2τ. Putting these together with eqa. 40a-f implies, in addition to eq. 34, that

$$T_{NS}(\rho_i,we^{2\pi i}) = (\frac{\tau}{i})^{-D/2}\ \tau^{-N}\ \hat{T}_R(\rho_i{}',w') \tag{42a}$$

$$T_R(\rho_i, w) = (\tfrac{\tau}{i})^{-D/2} \, \tau^{-N} \, \hat{T}_{NS}(\rho_i', w'e^{2\pi i}) \tag{42b}$$

$$T_{NS}(\rho_i, -w) = (\tfrac{\tau}{i})^{-D/2} \, (2\tau)^{-N} \, \hat{T}_{NS}(\rho_i'', -w''e^{2\pi i}) \tag{42c}$$

$$T_{NS}(\rho_i, -we^{2\pi i}) = (\tfrac{\tau}{i})^{-D/2} \, (2\tau)^{-N} \, \hat{T}_{NS}(\rho_i'', -w'') \tag{42d}$$

$$T_R(\rho_i, -w) = (\tfrac{\tau}{i})^{-D/2} (2\tau)^{-N} \, \hat{T}_R(\rho_i'', -w'') \tag{42e}$$

Note that the $(\tau/i)^{-D/2}$ factor is independent of twists or G parity projections since it arises from the zeroth mode of the bosonic oscillators. Elsewhere the appearance of -w in the Mobius loop correlations leads to extra factors of 2. In addition to eq. 35, the Neveu Schwarz and Ramond Partition functions satisfy

$$F_{NS}(we^{2\pi i}) = -\,(\tfrac{\tau}{i})^{D/2-1} \, F_R(w') \tag{43a}$$

$$F_R(w) = -\left(\frac{\tau}{i}\right)^{D/2-1} F_{NS}(w'e^{2\pi i}) \tag{43b}$$

$$F_{NS}(-w) = \left(\frac{2\tau}{i}\right)^{D/2-1} F_{NS}(-w''e^{2\pi i}) \tag{43c}$$

$$F_{NS}(-we^{2\pi i}) = \left(\frac{2\tau}{i}\right)^{D/2-1} F_{NS}(-w'') \tag{43d}$$

$$F_R(-w) = \left(\frac{2\tau}{i}\right)^{D/2-1} F_R(-w'') \tag{43e}$$

Extended to include double primed variables, eq. 36 reads

$$d\Omega = \tau^{N+1}\, d\Omega' = (2\tau)^{N+1}\, d\Omega'' \tag{44}$$

Inclusion of Mobius loops with multiple twists merely fills in the integration range of the ρ'' variables to match that of the ρ' variables. Thus when we substitute eqs. 34, 35, 42a-e, 43a-e, and 44 into the sum of eqs. 15 plus 20, we may drop all primes and double primes. Then labeling the ρ_i' variables by z_i

$$L = L_B + L_F$$

$$= i\frac{g^N}{2} \; d\Omega \; \{(\mathrm{Tr}\lambda^\circ)[F_{NS}(w)\hat{T}_{NS}(z_i,w) + F_{NS}(we^{2\pi i})\hat{T}_{NS}(z_i,we^{2\pi i})$$

$$-F_R(w)\hat{T}_R(z_i,w)]$$

$$-\;2^{D/2}[F_{NS}(-w)\hat{T}_{NS}(z_i,-w) + F_{NS}(-we^{2\pi i})\hat{T}_{NS}(z_i,-we^{2\pi i})$$

$$-F_R(-w)\hat{T}_R(z_i,-w)]\} \qquad (45)$$

We may now use Jacobi's famous "aequatio identica" relating, in effect, the Ramond and Neveu Schwarz partition functions.

$$F_R(w) = F_{NS}(w) + F_{NS}(we^{2\pi i}) \qquad (46)$$

The one loop N point function then becomes

$$L = \frac{i}{2}\, g^N \quad d\Omega \; \{(\mathrm{Tr}\lambda^\circ)[F_{NS}(w) \; (\hat{T}_{NS}(z_i,w) - \hat{T}_R(z_i,w))$$

$$+ F_{NS}(we^{2\pi i})(\hat{T}_{NS}(z_i,we^{2\pi i}) - \hat{T}_R(z_i,w))]$$

$$-\;2^{D/2}[F_{NS}(-w)(\hat{T}_{NS}(z_i,-w) - \hat{T}_R(z_i,-w))$$

$$+ F_{NS}(-we^{2\pi i})(\hat{T}_{NS}(z_i,-we^{2\pi i})-\hat{T}_R(z_i,-w))]\} \qquad (47)$$

Since the potential divergence in 47 arises from the $\frac{dw}{w}$

behavior of $d\Omega$, in order to prove finiteness one may discard terms in the curly bracket which vanish as $w \to 0$. Note that $w \to 0$ here is the point in the integration range where the original w of eqs. 11 through 22 approaches 1. The necessary expansions are

$$F_{NS}(w) = w^{-1/2} + 8 + O(w) \tag{48}$$

$$\hat{T}_{NS}(z_i, w) = \hat{T}_{NS}(z_i, 0) + w^{1/2}\, \hat{T}'_{NS}(z_i, 0) + O(w) \tag{49}$$

$$\hat{T}_R(z_i, w) = \hat{T}_R(z_i, 0) + O(w) \tag{50}$$

We have then

$$L = i\, g^N \int d\Omega\, \{((\mathrm{Tr}\lambda^\circ) - 2^{D/2})[8(\hat{T}_{NS}(z_i, 0) - \hat{T}_R(z_i, 0)) + \hat{T}'_{NS}(z_i, 0)] + O(w)\} \tag{51}$$

Since D = 10 in the superstring theory, all the N point functions are finite for $\mathrm{Tr}\, \lambda^\circ = 32$, ie for an SO(32) internal symmetry. From eq. 47 it is straightforward to recover the results of the light cone gauge calculations for small N. Since only the difference of the products of bosonic and fermionic vertices (eq 33) appears in eq. 47 there is no contribution from terms where only the $\zeta_i \cdot P_\mu$

pieces of V_B and V_F are taken. The vanishing of the two point function follows then from the relation

$$F_{NS}(w)[\chi^2(z,w) - \chi_0^2\,(z,w)] + (w \rightarrow we^{2\pi i}) = 0 \qquad (52)$$

Similarly the vanishing of the one loop 3 point function follows from the relation

$$F_{NS}(w)\ [\chi(z_2/z_1,w)\chi(z_3/z_2,w)\chi(z_3/z_1,w)$$
$$- \chi_0(z_2/z_1,w)\ \chi_0(z_3/z_2,w)\chi_0(z_3/z_1,w)]$$
$$+(w \rightarrow we^{2\pi i}) = 0 \qquad (53)$$

A similar relation among the x's also drastically simplifies the four point loop. We note that the vanishing of the one loop correction to the two and three point functions is independent of the internal symmetry group.

The techniques used here also greatly simplify the analysis of one loop graphs in the closed string models. Such results will appear in a later, more detailed article.

After completing this work, we heard that similar results have been obtained by Frampton, Moxhay, and Ng[5].

Much of the work discussed herein was completed at the University of Indiana and was supported in part by the

Department of Energy. Discussion of these results with participants at the Superstring Theory Workshops at the Lewes Center for Physics and the Santa Barbara ITP is gratefully acknowledged.

References

1. For reviews see J.H. Schwarz, Phys. Rep. 89, 223, 1982
   M. B. Green Surveys in High Energy Physics 3, 127, 1983.
2. M. B. Green and J. H. Schwarz Phy. Lett. 151B 21 1985.
3. P. Ramond Phy. Rev. D3 2415 1971.
   A. Neveu and J. H. Schwarz Nucl. Phys. B31 86 (1971).
   F. Gliozzi, J. Scherk and D. Olive Nucl. Phys. B122 253 (1977).
4. L. Clavelli and J. Shapiro Nucl. Phys. B57 490 1973.
5. Talk by P. Frampton. at Lewes Center for Physics Superstring Workshop July, 1985. These proceedings.

Appendix

The following transformation properties of Jacobi theta functions are used extensively in obtaining the results of this paper. The transformations to be considered are

$$\tau' = -\frac{1}{\tau} \qquad \nu' = \frac{\nu}{\tau} \tag{A.1}$$

$$\tau'' = -\frac{1}{4\tau} \qquad \nu'' = \frac{\nu}{2\tau} \tag{A.2}$$

The theta functions satisfy

$$\theta_1(\nu|\tau) = \frac{i\, e^{-i\pi\nu^2/\tau}}{\sqrt{\tau/i}} \quad \theta_1(\nu'|\tau') \tag{A.3}$$

$$\theta_2(\nu|\tau) = \frac{e^{-i\pi\nu^2/\tau}}{\sqrt{\tau/i}} \quad \theta_4(\nu'|\tau') \tag{A.4}$$

$$\theta_3(\nu|\tau) = \frac{e^{-i\pi\nu^2/\tau}}{\sqrt{\tau/i}} \quad \theta_3(\nu'|\tau') \tag{A.5}$$

$$\theta_4(\nu|\tau) = \frac{e^{-i\pi\nu^2/\tau}}{\sqrt{\tau/i}} \quad \theta_2(\nu'|\tau') \tag{A.6}$$

$$\theta_1'(0|\tau) = (\tau/i)^{-3/2} \quad \theta_1'(0|\tau') \tag{A.7}$$

$$\theta_1(\nu|\tau + 1) = e^{i\pi/4} \quad \theta_1(\nu|\tau) \tag{A.8}$$

$$\theta_2(\nu|\tau + 1) = e^{i\pi/4} \quad \theta_2(\nu|\tau) \tag{A.9}$$

$$\theta_3(\nu|\tau + 1) = \theta_4(\nu|\tau) \tag{A.10}$$

$$\theta_4(\nu|\tau + 1) = \theta_3(\nu|\tau) \tag{A.11}$$

$$\theta_1(\nu|\tau + 1/2) = \frac{ie^{-i\pi\nu^2/\tau}}{\sqrt{2\tau/i}} \theta_1(\nu''|\tau'' + 1/2) \tag{A.12}$$

$$\theta_2(\nu|\tau + 1/2) = \frac{e^{-i\pi\nu^2/\tau}}{\sqrt{2\tau/i}} \theta_2(\nu''|\tau'' + 1/2) \tag{A.13}$$

$$\theta_3(\nu|\tau + 1/2) = \frac{e^{i\pi/4}\ e^{-i\pi\nu^2/\tau}}{\sqrt{2\tau/i}} \theta_4(\nu''|\tau'' + 1/2) \tag{A.14}$$

$$\theta_4(\nu|\tau + 1/2) = \frac{e^{-i\pi/4}\ e^{-i\pi\nu^2/\tau}}{\sqrt{2\tau/i}} \theta_3(\nu''|\tau'' + 1/2) \tag{A.15}$$

$$\theta_1'(0|\tau + 1/2) = (2\tau/i)^{-3/2} \ \theta_1'(0|\tau'' + 1/2) \tag{A.16}$$

The relations between the elementary correlations and the Jacobi theta functions are for $\nu = \frac{\ln(\rho_j/\rho_i)}{2\pi i}$ and $\tau = \ln w/2\pi i$

$$\langle 0|Q_\mu(\rho_i,d,\bar{d})Q_\nu(\rho_j,d,\bar{d})|0\rangle$$

$$= g_{\mu\nu}\left[-\frac{1}{\epsilon^2 \ln w} - \frac{\ln w}{12} - \ln\Psi(\rho_j/\rho_i,w)\right] \tag{A.17}$$

$$\Psi(\rho_j/\rho_i,w) = -2\pi i e^{i\pi\nu^2/\tau}\frac{\theta_1(\nu|\tau)}{\theta_1'(0|\tau)} \tag{A.18}$$

$$\langle 0|\hat{Q}_\mu(\rho_i,d,\bar{d})\hat{Q}_\nu(\rho_j,d,\bar{d})|0\rangle$$

$$= g_{\mu\nu}\left[-\frac{1}{2\epsilon} - \ln\hat{\Psi}(\rho_j/\rho_i,w)\right] \tag{A.19}$$

$$\hat{\Psi}(\rho_j/\rho_i,w) = -2\pi i\frac{\theta_1(\nu|\tau)}{\theta_1'(0|\tau)} \tag{A.20}$$

Equation 28a then follows immediately from the theta function transformation properties. Equations 28b, c follow by differentiation using the definition eq 13.

The vev's of the Neveu Schwarz and Ramond fields are

$$\langle 0|H_\mu(\rho_i,d,\bar{d})H_\nu(\rho_j,d,\bar{d})|0\rangle$$

$$= \frac{i}{2}\theta_2(0|\tau)\theta_4(0|\tau)\frac{\theta_3(\nu|\tau)}{\theta_1(\nu|\tau)}g_{\mu\nu} \tag{A.21}$$

$$\langle 0|\Gamma_\mu(\rho_i,d,\bar{d})\Gamma_\nu(\rho_j,d,\bar{d})|0\rangle\ (-2)^{-1}$$

$$= \frac{i}{2}\,\theta_3(0|\tau)\theta_4(0|\tau)\ \frac{\theta_2(\nu|\tau)}{\theta_1(\nu|\tau)}\ g_{\mu\nu} \tag{A.22}$$

leading, therefore, in a straightforward way to eqs 40a-f.

The modified partition functions of eqs 16 and 21 are

$$F_{NS}(w) = w^{-1/2} \prod_{n=1}^{\infty}\left(\frac{1+w^{n-1/2}}{1-w^n}\right)^{D-2}$$

$$= \left[2\pi\ \frac{\theta_3(0|\tau)}{\theta_1{}'(0|\tau)}\right]^{\frac{D-2}{2}} \tag{A.23}$$

$$F_R(w) = 2^{\frac{D-2}{2}} \prod_{n=1}^{\infty}\left(\frac{1+w^n}{1-w^n}\right)^{D-2}$$

$$= \left[2\pi\ \frac{\theta_2(0|\tau)}{\theta_1{}'(0|\tau)}\right]^{\frac{D-2}{2}} \tag{A.24}$$

The Jacobi relation leading to eq 46 is

$$\theta_2(0|\tau)^4 = \theta_3(0|\tau)^4 - \theta_4(0|\tau)^4 \tag{A.25}$$

The situation in the case of multiple twists deserves some comment. If there are an odd number (>1) of twists in the loop, one or more of the ρ_i in eq 17 become negative in addition to $w \rightarrow -w$. The sum of all possible odd numbers of twists results in each ρ_i i = 2 ...N being integrated over negative as well as positive values. We have the relations:

$$\Psi(\rho_j/\rho_i,w) = \tau\,\hat{\Psi}(\rho_j'/\rho_i',w') \tag{A.26}$$

$$\Psi(\rho_j/\rho_i,-w) = (2\tau)\,\hat{\Psi}(\rho_j''/\rho_i'',w'') \tag{A.27}$$

As can be seen from eqs 23 and 25, the ρ_i' are integrated around the full unit circle whereas the ρ_i'' are integrated around half of the unit circle. Terms with ρ_j replaced by $-\rho_j$ will depend on $\Psi(-\rho_j/\rho_i,-w)$. But

$$\Psi(-\rho_j/\rho_i,-w) = (2\tau)\,\hat{\Psi}(e^{i\pi}\rho_j''/\rho_i'',-w'') \tag{A.28}$$

The effect is to extend the ρ_j'' integration around the full unit circle.

A slight complication occurs when one considers the Neveu Schwarz and Ramond correlations of eqs 38 and 39. Whereas

$$\chi(\rho_j/\rho_i,-w) = \frac{1}{2\tau}\,\chi(\rho_j''/\rho_i,-w''\,e^{2\pi i}) \tag{A.29}$$

and

$$\chi(-\rho_j/\rho_i,-w) = \frac{1}{2\tau}\,\chi(e^{i\pi}\rho_j''/\rho_i,-w''\,e^{2\pi i}) \tag{A.30}$$

consistent with the simple extension of the ρ'' integration range, one finds

$$\chi(\rho_j/\rho_i, -we^{2\pi i}) = \frac{1}{2\tau}\,\chi(\rho_j''/\rho_i'', -w'') \tag{A.31}$$

but on the other hand

$$\chi(-\rho_j/\rho_i, -we^{2\pi i}) = -\,\frac{1}{2\tau}\,\chi(e^{i\pi}\rho_j''/\rho_i'', -w'') \tag{A.32}$$

The extra minus sign in A.32 together with a similar sign change in the fermionic correlation

$$\chi_0(-\rho_j/\rho_i, -w) = -\,\frac{1}{2\tau}\,\chi_0(e^{i\pi}\rho_j''/\rho_i'', -w'') \tag{A.33}$$

might lead one to worry until one notes that each ρ_i occurs in two χ's or χ_0's. The extra minus sign, therefore, always occurs squared. The net effect of the multiple twists is, therefore, to extend the ρ_i'' integrations over the full circle to match the integration range of the ρ_i'. This justifies our discarding of the distinction between ρ_i' and ρ_i'' in eq 45.

TOWARD A COVARIANT STRING FIELD THEORY

Stuart Raby, Richard Slansky, and Geoffrey West

Theoretical Division, T-8, MS B285
Los Alamos National Laboratory
Los Alamos, New Mexico 87545 USA

ABSTRACT

We identify the physical degrees of freedom in the string field using the equations of motion, gauge conditions and residual gauge invariance of the system. We also present a first step towards obtaining a covariant string field theory.

INTRODUCTION

We want to discuss string field theory which we shall distinguish from the particle theory of the string. Moreover, we want to describe a string field theory which is both Lorentz invariant and covariant under local gauge transformations which we shall soon identify.

The point particle has canonical coordinates $x^{\mu}(\tau)$ which describe its motion along a world line (see fig. 1) in space-time. The parameter τ labels the position along the world line. The action

$$S = -m\int d\tau \sqrt{-\left(\frac{dx}{d\tau}\right)^2}$$

is invariant under arbitrary τ reparametrizations $\tau \to \tau + \xi(\tau)$. As a

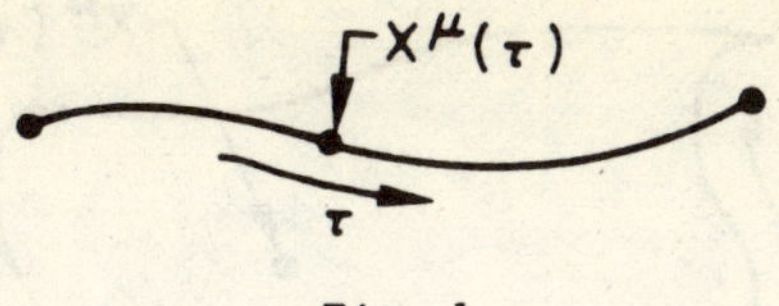

Fig. 1

result, we obtain the primary constraint

$$p^2 + m^2 = 0 \quad .$$

The field theory of the point particle satisfies the equations of motion

$$(\Box - m^2)\phi(x) = 0 \quad ,$$

where $\phi(x)$ is the real Klein-Gordon field which is a function of the space-time coordinate x^μ. The second quantized field ϕ creates and annihilates particles with the quantum numbers of the point particle. Note that the equations of motion are obtained from the primary constraints of the point particle theory, with the identification $p_\mu \to -i\partial/\partial x^\mu$. The reparametrization invariance of the point particle theory is not apparent in the corresponding field theory.

The particle theory of a one-dimensional string is described by the canonical coordinates

$$X^\mu(\tau,\sigma) \quad ,$$

where σ parametrizes the position along the string (at fixed τ) and τ parametrizes its motion in space-time (see fig. 2). The action for the string

$$S = \int d\tau d\sigma \sqrt{\left(\frac{\partial X^\mu}{\partial\sigma}\frac{\partial X_\mu}{\partial\tau}\right)^2 - \left(\frac{\partial X^\mu}{\partial\sigma}\frac{\partial X_\mu}{\partial\sigma}\right)\left(\frac{\partial X^\nu}{\partial\tau}\frac{\partial X_\nu}{\partial\tau}\right)}$$

is invariant under arbitrary σ,τ reparametrizations

$$\sigma \to \sigma + \xi_\sigma(\tau,\sigma) \quad ,$$

$$\tau \to \tau + \xi_\tau(\tau,\sigma) \quad .$$

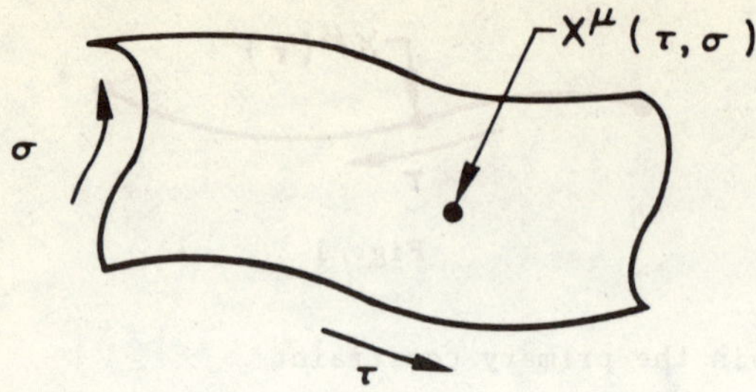

Fig. 2

As a result we obtain the primary constraints $L_n = 0$ where L_n are the generators of the Virasoro algebra. Free string field theory is then described by the equations

$$(L_0 - 1)\,\Phi\,[X(\sigma)] = 0 \quad ,$$

$$L_n\,\Phi\,[X(\sigma)] = 0 \quad , \quad n > 0 \quad ,$$

where the string field Φ is a hermitian functional of the coordinates $X^\mu(\sigma)$. Note we have used only half of the primary constraints. This is because our representation of the Virasoro algebra includes a central term which does not permit us to satisfy $L_n\Phi = 0$ for all n. Open and closed strings include gauge fields. One of the questions we shall address is, in what way are the local gauge symmetries of the string field theory related to the two-dimensional reparametrization invariance of the string particle theory?

Why are we interested in string theories? The particle theory of the string provides an algorithm for calculating scattering amplitudes for the states of the string. It has been shown[1)] that the ten dimensional superstring is 1) one-loop finite, 2) anomaly free, and 3) contains enough states for describing a truly "unified field theory" of nature. The particle theory of the string is quite complicated. The field theory of the string does not promise to be any simpler. Why then, do we want to construct a string field theory?

a) Bosonic strings contain tachyons. In a scalar field theory, for example, a tachyonic state is usually a symptom that one is quantizing the theory about the wrong ground state (see fig. 3). To quantize about the correct ground state, one minimizes the potential and then shifts the field by the appropriate amount. In the particle theory the background is fixed and thus finding a better ground state (should one exist) requires summing an infinite number of perturbation theory diagrams.

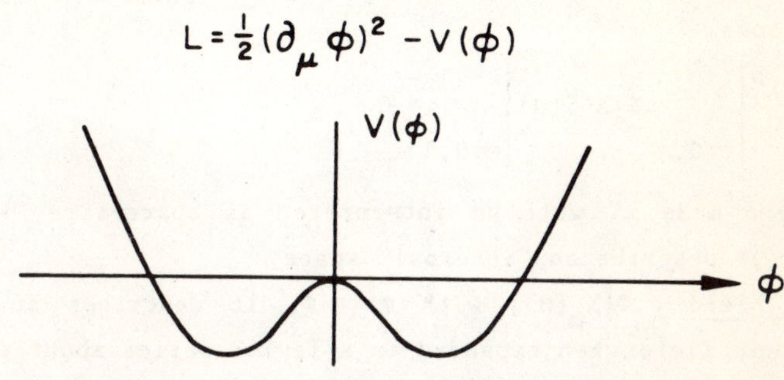

Fig. 3

b) The closed string contains a massless symmetric tensor state - a "graviton".

i) classical solutions of the closed string field theory may provide some insight as to the difference between string gravity and Einstein gravity at short distances.

ii) A ten or 26 dimensional string must compactify to give us flat 4 dimensional Minkowski space. In Kaluza-Klein theories one looks for classical solutions to the field equations on a compact manifold. One would like to understand the vacuum geometry for the string.

iii) Finally and most importantly, it is clear that the standard general coordinate invariance of Einstein gravity is not a symmetry of the string. For example, there does not exist a local 4 point interaction for the graviton in string theories (unless one takes the zero slope limit). What is the "symmetry principle" which replaces general coordinate invariance as a symmetry of nature?

OPEN STRING

Coordinates:

$$X^\mu(\sigma) = x^\mu + 2\sum_{\ell=1}^{\infty} \cos\ell\sigma\, x^\mu_\ell \quad ,$$

where $\mu = 1,\ldots,D$; $0 \leqq \sigma \leqq \pi$. The coordinates satisfy the boundary conditions

$$\left.\frac{dX^\mu}{d\sigma}\right|_{\sigma=0,\pi} \equiv \left. X'^\mu(\sigma)\right|_{\sigma=0,\pi} = 0 \quad .$$

The zero mode x^μ will be interpreted as space-time, while x^μ_ℓ, $\ell=1,\ldots,\infty$ describe an "internal" space.

String Field: $\Phi[X_\mu(\sigma)]$ with $\Phi^+ = \Phi$. It describes an infinite component field when expanded in a Taylor series about the point x^μ.

Differential operators on functionals:

$$\frac{\delta}{\delta X^\mu(\sigma)} = \frac{1}{\pi}\sum_{\ell=0}^{\infty} \cos\ell\sigma \frac{\delta}{\delta x^\mu_\ell}$$

with ordinary space-time differentiation given by

$$\partial_\mu \equiv \frac{\delta}{\delta x^\mu_0} \quad .$$

Virasoro operators: We want to obtain a representation of L_n acting on string fields. In the first quantized string the L_n's are given by

$$L_n = \frac{1}{2} \quad : \sum_{\ell=-\infty}^{+\infty} a^\dagger_\ell \cdot a_{\ell+n} \quad :$$

where the creation and annihilation operators $a^{\mu\dagger}_n$, a^μ_n satisfy

i) $\qquad a^{\dagger\mu}_n \equiv a^\mu_{-n} \qquad , \quad n \geqq 1 \;,$

ii) $\quad [a^\mu_n, a^{\dagger\nu}_m] = n\delta_{nm}\eta^{\mu\nu} \;.$

The brackets : : denote normal ordering. The L_n's satisfy

i) $\quad L^\dagger_n = L_{-n} \quad ,$

ii) $[L_n, L_m] = (n-m)\, L_{n+m} + \frac{D}{12}\,(n^3-n)\delta_{n+m,0}$.

A representation for the a's is obtained as follows. Define

$$\left.\begin{array}{l} P^{\mu}(\sigma) = i\dfrac{\delta}{\delta X_{\mu}(\sigma)} + X'^{\mu}(\sigma) \\ \\ \text{and} \\ \\ P^{\mu}(-\sigma) = i\dfrac{\delta}{\delta X_{\mu}(\sigma)} - X'^{\mu}(\sigma) \end{array}\right\} \quad 0 \leqq \sigma \leqq \pi$$

Then

$$a_n^{\mu} \equiv \frac{1}{\sqrt{4\pi}} \int_{-\pi}^{\pi} d\sigma\, e^{-in\sigma}\, P^{\mu}(\sigma) \quad .$$

We have

$$a_0^{\mu} \equiv \frac{i}{\sqrt{\pi}}\,\partial^{\mu}$$

$$a_n^{\mu} = \frac{i}{\sqrt{4\pi}}\left(\frac{\delta}{\delta x_{n\mu}} + 2\pi n\, x_n^{\mu}\right) \quad , \quad n \neq 0 \quad .$$

The Virasoro operators are then defined as operators on string fields using the above definition of the a's. For later use we write explicitly L_0, L_{-1}, L_{-2}.

$$L_0 = \frac{1}{2\pi}\,(2\pi N - \Box) \quad ; \quad N = \sum_{\ell=1}^{\infty} a_{\ell}^{\dagger} a_{\ell} \quad ,$$

$$L_{-1} = \frac{i}{\sqrt{\pi}}\, a_1^{\dagger} \cdot \partial + \sum_{\ell=1}^{\infty} a_{\ell+1}^{\dagger} \cdot a_{\ell} \quad ,$$

$$L_{-2} = \frac{i}{\sqrt{\pi}}\, a_2^{\dagger} \cdot \partial + \frac{1}{2}\, a_1^{\dagger} \cdot a_1^{\dagger} + \sum_{\ell=1}^{\infty} a_{\ell+2}^{\dagger} \cdot a_{\ell} \quad .$$

<u>Physical degrees of freedom</u> (PDOF):

We shall now identify the PDOF contained in Φ. We use, A) the field equations, B) the gauge conditions, and C) the residual local gauge invariance of A) and B).

A) $\Theta_0 \Phi = 0 \quad ; \quad \Theta_0 \equiv L_0 - 1 \quad ,$

B) $L_n \Phi = 0 \quad ; \quad n \geqq 1 \quad .$

A) and B) are invariant under the local transformation

C) $\delta\Phi = \sum_{m=1}^{\infty} L_{-m}\Lambda_m$

where

$\Lambda_m \equiv \Lambda_m[X(\sigma)]$ satisfy

$L_{+n}\Lambda_m = -(2n+m)\,\Lambda_{n+m} \quad ,$ (a)

$(L_0 + m-1)\Lambda_m = 0 \quad .$ (b)

This symmetry only exists for D = 26. As a result of (a) we see that

(1) Λ_n for $n \neq 0$ can be expressed in terms of Λ_1, i.e.,

$$\Lambda_n = \frac{2(-1)^{n-1}}{(n+1)!} L_1^{\,n-1} \Lambda_1 \quad ,$$

and

(2) Λ_1 is constrained to satisfy

$$L_0\Lambda_1 = 0$$

$$L_{+2}\Lambda_1 = -\frac{5}{12} L_{+1}^2\Lambda_1 \quad , \text{etc.}$$

(3) Not all the independent parameters contained in Λ_1 are relevant, i.e., the above constraints on Λ_1 are invariant under the transformation

$$\delta\Lambda_1 = \sum_{m=2}^{\infty} L_{-m}\,\chi_m$$

where the $\chi_m \equiv \chi_m\,[X(\sigma)]$ satisfy

$L_{+n}\chi_m = -(2n+m)\,\chi_{n+m} \quad ,$

$(L_0 + m)\,\chi_m = 0 \quad ,$

and

$$\delta^2\Phi = \sum_{m=1}^{\infty} L_{-m}\delta\Lambda_m \equiv 0 \quad .$$

The last equation says that not all the independent components of Λ_1 will change Φ, i.e., they are not all relevant. The above process continues. We have checked that using the results of A), B), and C), we find a one-to-one correspondence between the states of the first quantized string in the light cone gauge and the PDOF in Φ up to mass level 5. Let us discuss a few simple examples.

The solution to A) is given by the following construction. Define the ground state functional $\Phi_0[X(\sigma)]$ such that

$$a_n\Phi_0 = 0 \quad , \quad n \geqq 0 \quad .$$

Φ_0 is a gaussian functional in the internal coordinates x_ℓ^μ , $\ell \neq 0$. Then a solution to A) is

$$\Phi[X(\sigma)] = \{\phi(x) + A_\mu^\ell(x)\, a_\ell^{\dagger\mu} + \tfrac{1}{2}h_{\mu\nu}^{\ell m}(x)\, a_\ell^{\dagger\mu} a_m^{\dagger\nu} + \dots \}\, \Phi_0 \quad ,$$

$$\ell, m = 1, \dots \infty \quad ,$$

where the coefficient functions satisfy the field equations

$$(\Box + 2\pi)\,\phi = 0 \quad ,$$

$$\Box A_\mu^1 = 0 \quad ,$$

$$(\Box - 2\pi)\, A_\mu^2 = 0 \quad ,$$

$$(\Box - 2\pi)\, h_{\mu\nu}^{11} = 0 \quad ,$$

etc.

ϕ is thus a tachyon with $m^2 = -2\pi$, A_μ^1 is a Maxwell field [the first mass level], and A_μ^2 and $h_{\mu\nu}^{11}$ have $m^2 = 2\pi$ [the second mass level]. The field equations A) enable us to identify an infinite set of local fields and mass levels. However, not all the fields are independent. We use B) + C) to identify the independent PDOF. Consider the first 2 mass levels.

$\underline{m^2 = 0}$

A) $\Theta_0\Phi = 0$ gives $\Box A^1_\mu = 0$,

B) $L_1\Phi = 0$ gives $\partial \cdot A^1 = 0$,

i.e., the Lorentz gauge condition. It is for this reason we have identified the equations B) as gauge conditions. Finally

C) $\delta\Phi = L_{-1}\Lambda_1 + \ldots$ gives $\delta A^1_\mu = \partial_\mu \lambda_1$,

where

$$\Lambda_1[x(\sigma)] = [\lambda_1(x) + \lambda^1_\mu(x)\, a_1^{\dagger\mu} + \ldots]\,\Phi_0 \quad ,$$

and the constraint $L_0\Lambda_1 = 0$ implies $\Box\lambda_1(x) = 0$. Thus, the Maxwell field A^1_μ has D components, of which only D-2 are physical since B) says one field is dependent and C) says that another can be gauged away. Note that to obtain a gauge invariant system we must remove the constraint on the gauge parameter, i.e., we would like

$$\delta A^1_\mu = \partial_\mu \lambda_1 \text{ for arbitrary } \lambda_1(x) \quad .$$

At the second mass level $\underline{(M^2 = 2\pi)}$, we have

A) $\Theta_0\Phi = 0$ gives $(\Box - 2\pi)\, A^2_\mu = 0$,

$$(\Box - 2\pi) h^{11}_{\mu\nu} = 0 \quad ,$$

or

$D + \frac{D(D+1)}{2}$ fields. They satisfy the gauge conditions

B) $L_1\Phi = 0$ gives $2A^2_\mu + \frac{i}{\sqrt{\pi}}\, \partial^\nu h^{11}_{\mu\nu} = 0$

$L_2\Phi = 0$ gives $\frac{2i}{\sqrt{\pi}}\, \partial\cdot A^2 + \frac{1}{2} h^{11} = 0$

with $h^{11} \equiv h^{11\mu}_\mu$, or

D + 1 gauge conditions. Finally we have the invariance

C) $\delta\Phi = (L_{-1} - \frac{1}{3} L_{-2}L_{1})\Lambda_{1}$ which gives

$$\delta A^{2}_{\mu} = -\lambda^{1}_{\mu} + \frac{1}{3\pi}\,\partial_{\mu}\partial\cdot\lambda^{1} ,$$

$$\delta h^{11}_{\mu\nu} = \frac{-i}{\sqrt{\pi}}\,[\partial_{\mu}\lambda^{1}_{\nu} + \partial_{\nu}\lambda^{1}_{\mu} - \frac{1}{3}\eta_{\mu\nu}\partial\cdot\lambda^{1}]$$

with the constraint $L_{0}\Lambda_{1} = 0$ giving $(\Box-2\pi)\lambda^{1}_{\mu} = 0$, or D independent gauge functions. We thus find

$$D + \frac{D(D+1)}{2} - (D + 1) - D = \frac{(D+1)(D-2)}{2}$$

independent PDOF. This is the correct count to describe a massive, traceless, symmetric tensor.

FIRST STEP TOWARD A COVARIANT FIELD THEORY

We shall obtain a solution to the following demands.

1) Remove the constraint on the gauge functional i.e., find a new equation of motion $\Theta\Phi = 0$ invariant under the local transformation

$$\delta\Phi = \sum_{m=1}^{\infty} L_{-m}\Lambda_{m}$$

without the constraint

$$(L_{0} + m - 1)\,\Lambda_{m} = 0 \quad .$$

2) The equations of motion are local in terms of the component fields.

3) No auxiliary fields are needed.

4) The gauge conditions $L_{n}\Phi = 0$ are obtained as a gauge choice.

We find a solution for D = 26

$$\Theta = L_{0}-1 - \sum_{n=1}^{\infty} \frac{L_{-n}L_{n}}{2n} \quad .$$

As an example the equations of motion $\Theta\Phi = 0$ gives for the Maxwell field

$$\partial^\mu F_{\mu\nu} = 0 \text{ with } F_{\mu\nu} = \partial_\mu A^1_\nu - \partial_\nu A^1_\mu \ .$$

This is invariant under the transformation $\delta\Phi = L_{-1}\Lambda_1 + \ldots$ which gives $\delta A^1_\mu = \partial_\mu \lambda_1$ with arbitrary $\lambda_1(x)$. We note that the original gauge conditions $L_n\Phi = 0, n \geqq 1$ may be obtained by a choice of Λ_1. It is however necessary to use the equations of motion $\Theta\Phi = 0$.

Closed String:

A closed string is defined by the coordinates

$$X^\mu(\sigma) = \sum_{\ell=-\infty}^{+\infty} e^{i\ell\sigma} x^\mu_\ell \ , \ -\pi \leqq \sigma \leqq \pi \ ,$$

with

$$x^\mu_{-\ell} \equiv x^{\mu *}_\ell \ , \quad \mu = 1, \ldots, D$$

(see fig. 4). $X^\mu(\sigma)$ describes a closed loop in space-time. Once again space-time is associated with the normal mode $x^\mu_0 \equiv x^\mu$ and x^μ_ℓ, $\ell \neq 0$ are "internal spaces."

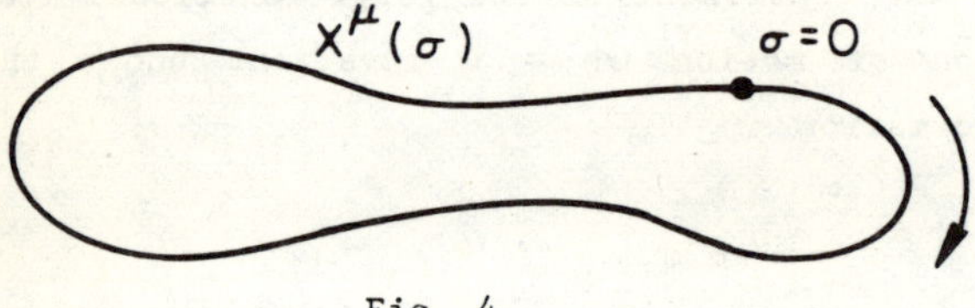

Fig. 4

The <u>string field</u> $\Phi[X_\mu(\sigma)]$ is a real functional of the coordinates; $\Phi = \Phi^\dagger$. Since the origin of the loop has no physical significance we shall require $\Phi[X]$ to be invariant under redefinitions of the origin, i.e. $\Phi[X(\sigma+c)] = \Phi[X(\sigma)]$ where c is an arbitrary constant.

Once again we shall define <u>differential operators</u> on closed string functionals and then use these operators to construct representations of the Virasoro algebra. We define the differential

$$\frac{\delta}{\delta X^\mu(\sigma)} = \frac{1}{2\pi} \sum_{\ell=-\infty}^{+\infty} e^{-i\ell\sigma} \frac{\delta}{\delta x^\mu_\ell}$$

with space-time differentiation given by

$$\partial_\mu \equiv \frac{\delta}{\delta x_0^\mu} \quad .$$

We then define two sets of creation and annihilation operators

$$P^\mu(\sigma) = i\frac{\delta}{\delta X_\mu(\sigma)} + X'^\mu(\sigma) = \frac{1}{\sqrt{\pi}} \sum_{\ell=-\infty}^{+\infty} e^{i\ell\sigma} a_\ell^\mu \quad ,$$

$$\bar{P}^\mu(\sigma) = i\frac{\delta}{\delta X_\mu(\sigma)} - X'^\mu(\sigma) = \frac{1}{\sqrt{\pi}} \sum_{\ell=-\infty}^{+\infty} e^{-i\ell\sigma} \bar{a}_\ell^\mu \quad ,$$

where in the particle theory a_ℓ^μ and $\bar{a}_\ell^\mu$ describe the states of right and left movers, respectively. Note that

$$a_0^\mu = \bar{a}_0^\mu = \frac{i}{\sqrt{4\pi}} \partial^\mu \quad .$$

We then obtain a Virasoro algebra for right (L_n) and left ($\bar{L}_n$) movers. We have

$$L_0 = \frac{1}{8\pi}(-\Box + 8\pi N) \quad ,$$

$$\bar{L}_0 = \frac{1}{8\pi}(-\Box + 8\pi N) .$$

The generator of constant σ-translations is given by $L_0 - \bar{L}_0$. We thus require Φ to satisfy

$$(L_0 - \bar{L}_0)\Phi = (N - \bar{N})\Phi = 0 \quad ,$$

where $N(\bar{N})$ are number operators for right (left) movers.

Let us now discuss the physical degrees of freedom contained in Φ. We use

A) the <u>field equations</u> $\Theta_0 \, \Phi = 0 \quad , \quad \Theta_0 = L_0 - 1$;

B) the <u>gauge conditions</u> $L_n \, \Phi = \bar{L}_n \, \Phi = 0 \quad , \quad n \geq 1$, and

C) the <u>residual gauge invariance</u> of A) and B)

$$\delta\Phi = L_{-1}\bar{\Lambda}_1 + \bar{L}_{-1}\Lambda_1 + L_{-1}\bar{L}_{-1}\Lambda_{11}$$

where the gauge functionals satisfy the constraints

$$L_0\Lambda_{11} = L_0\bar{\Lambda}_1 = \bar{L}_0\Lambda_1 = 0 \quad .$$

They also satisfy

$$(N-\bar{N})\bar{\Lambda}_1 = \bar{\Lambda}_1 \quad , \quad (N-\bar{N})\Lambda_1 = -\Lambda_1 \quad ,$$

$$(N-\bar{N})\Lambda_{11} = 0 \quad ,$$

$$L_1\bar{\Lambda}_1 = \bar{L}_1\bar{\Lambda}_1 = L_1\Lambda_1 = \bar{L}_1\Lambda_1 = 0 \quad .$$

Note that this is not the complete residual invariance of the theory, but only that piece which affects the zero mass level which we shall now discuss. Consider the solution to A) the field equation. We define the ground state functional Φ_0, satisfying

$$a^\mu_n \Phi_0 = \bar{a}^\mu_n \Phi_0 = 0 \text{ for } n \geqq 0 \quad .$$

Then

$$\Phi[X^\mu(\sigma)] = \{\phi(x) + h^\ell_{\mu\nu} \bar{a}^{\mu\dagger}_\ell a^{\nu\dagger}_\ell + \ldots\}\Phi_0 \quad ,$$

where the component fields satisfy $(\Box+8\pi)\phi(x) = 0$, a tachyon with $\text{mass}^2 = -8\pi$ and $\Box h^1_{\mu\nu} = 0$, a massless field with D^2 components; etc. Once again we can use the gauge conditions B) and residual gauge invariance C) to identify the PDOF. We find (at the zero mass level) that $h_{\mu\nu} \equiv 1/2(h^1_{\mu\nu} + h^1_{\nu\mu})$ describes a massless symmetric tensor field with $(D^2-3D)/2$ PDOF, i.e., a graviton and $A_{\mu\nu} \equiv 1/2(h^1_{\mu\nu} - h^1_{\nu\mu})$ describes a massless anti-symmetric tensor field with $(D^2-5D+6)/2$ PDOF, i.e., a Kalb-Ramond gauge field. Note that the massless dilaton which exists in the particle theory is missing.

We shall now obtain a new equation of motion given by A') $\Theta_1\Phi = 0$ where Φ now satisfies the gauge conditions B') $L_n\Phi = \bar{L}_n\Phi = 0 \quad n \geqq 2$ and A') and B') are invariant under the local gauge transformation

$$\delta\Phi = L_{-1}\bar{\Lambda}_1 + \bar{L}_{-1}\Lambda_1 + L_{-1}\bar{L}_{-1}\Lambda_{11}$$

with the constraints removed i.e., $L_0\Lambda_{11} \neq 0$, $L_0\bar{\Lambda}_1 \neq 0$, and $\bar{L}_0\Lambda_1 \neq 0$. We find

$$\Theta_1 = \frac{1}{2}[L_0 + \bar{L}_0 - 2 - L_{-1}L_1 - \bar{L}_{-1}\bar{L}_1 + L_{-1}\bar{L}_{-1}\left(\frac{1}{L_0+\bar{L}_0}\right)L_1\bar{L}_1] \quad .$$

The field equation in the massless sector is

$$\Box h^1_{\mu\nu} - \partial_\mu\partial^\lambda h^1_{\lambda\nu} - \partial_\nu\partial^\lambda h^1_{\mu\lambda} + \partial_\mu\partial_\nu \frac{1}{\Box} \partial^\lambda\partial^\zeta h^1_{\lambda\zeta} = 0 \quad .$$

Although it is non-local, by taking a trace, we obtain

$$\Box h^{1\mu}_{\mu} = \partial^\mu\partial^\nu h^1_{\mu\nu}$$

and thus it is equivalent to the local equation

$$\Box h^1_{\mu\nu} - \partial_\mu\partial^\lambda h^1_{\lambda\nu} - \partial_\nu\partial^\lambda h^1_{\mu\lambda} + \partial_\mu\partial_\nu h^{1\lambda}_{\lambda} = 0.$$

These are just the linearized equations of Einstein gravity for $h_{\mu\nu}$ and the gauge invariant equations for $A_{\mu\nu}$. Note this result is a necessary consequence of gauge symmetry $\delta\Phi$ (discussed above) which in component form implies

$$\delta h_{\mu\nu} = \partial_\mu\lambda_\nu - \partial_\nu\lambda_\mu \quad ,$$

$$\delta A_{\mu\nu} = \partial_\mu\lambda'_\nu - \partial_\nu\lambda'_\mu \quad ,$$

It is clear how to solve simultaneously the problem of the missing dilaton and the non-locality of the equations of motion. Consider the non-local linearized Einstein equations

$$\Box h_{\mu\nu} - \partial_\mu\partial^\lambda h_{\lambda\nu} - \partial_\nu\partial^\lambda h_{\mu\lambda} + \partial_\mu\partial_\nu \frac{1}{\Box} \partial^\lambda\partial^\zeta h_{\lambda\zeta} = 0 \quad .$$

This equation may be made local by introducing a new scalar field ϕ' satisfying

$$\Box h_{\mu\nu} - \partial_\mu\partial^\lambda h_{\lambda\nu} - \partial_\nu\partial^\lambda h_{\mu\lambda} + \partial_\mu\partial_\nu\phi' = 0 \quad ,$$

$$\Box\phi' = \partial^\lambda\partial^\zeta h_{\lambda\zeta} \quad .$$

Note that these equations are invariant under the transformations

$$\delta h_{\mu\nu} = \partial_\mu\lambda_\nu + \partial_\nu\lambda_\mu \quad ,$$

$$\delta h_\mu{}^\mu = 2\partial \cdot \lambda \quad ,$$

$$\delta\phi' = 2\partial \cdot \lambda \quad .$$

Thus, the linear combination $\phi' - h_\mu{}^\mu$ is gauge invariant and indeed one finds that it satisfies the equation of motion

$$\Box(\phi' - h^\mu_\mu) = 0 \quad .$$

The non-locality of the string field equations can then be remedied by the same procedure. Introduce a new functional $\Phi'[X^\mu(\sigma)] = \phi'(x)\Phi_0 + \ldots$, with the coupled field equations

$$(L_0 + \bar{L}_0 - 2 - L_{-1}L_1 - \bar{L}_{-1}\bar{L}_1)\Phi - L_{-1}\bar{L}_{-1}\Phi' = 0 \quad ,$$

$$(L_0 + \bar{L}_0)\Phi' = -L_1\bar{L}_1\Phi \quad .$$

These equations are invariant under the gauge transformations

$$\delta\Phi = L_{-1}\bar{\Lambda}_1 + \bar{L}_{-1}\Lambda_1 \quad ,$$

$$\delta\Phi' = \bar{L}_1\bar{\Lambda}_1 + L_1\Lambda_1 \quad ,$$

Interactions:

Up until now we have discussed free string field theory. Interactions may be introduced into the action with terms of the form (see fig. 5)

$$S_I = \int \prod_{i=1}^{3} dX_i^\mu(\sigma)\Phi[X_1]\Phi[X_2]\Phi[X_3]V[X_1,X_2,X_3] \quad .$$

We define an energy-momentum tensor functional

$$T[X_1(\sigma)] = \frac{\delta S_I}{\delta\Phi[X_1(\sigma)]} \quad ,$$

$$= \int \prod_{i=2}^{3} dX_i^\mu(\sigma)\Phi[X_2]\Phi[X_3]V[X_1,X_2,X_3] \quad .$$

S_I is constrained to satisfy the linearized gauge transformations. As a result, we find the generalized conservation law

$$L_1 T[X_1(\sigma)] = \bar{L}_1 T[X_1(\sigma)] = 0 \quad , \qquad \text{(on shell)} \quad .$$

It will be through the study of the interacting string that we may learn the non-linear nature of the symmetry. It is at this point we will discover whether the symmetry is, for example, a space-time symmetry or rather an internal symmetry of the string.

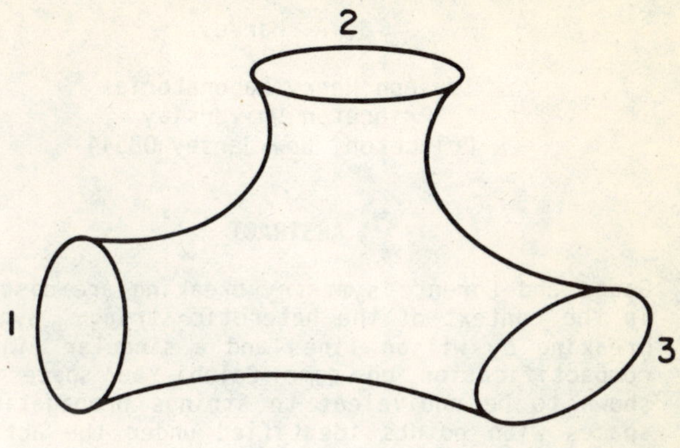

Fig. 5

TWISTING THE HETEROTIC STRING

J. A. Harvey*

Joseph Henry Laboratories
Princeton University
Princeton, New Jersey 08544

ABSTRACT

Gauge and Lorentz symmetry breaking are discussed in the context of the heterotic string. Symmetry breaking by Wilson lines and a singular limit of compactification on some Calabi-Yau spaces are shown to be equivalent to strings propagating on spaces with points identified under the action of discrete symmetries. It is also shown that certain twisted two dimensional heterotic string theories naturally incorporate the Frenkel-Lepowsky-Meurman construction of the monster goup F_1.

1. INTRODUCTION

String theories have the potential to give a truly unified description of all known interactions. Of course this unification comes only at the expense of an enormous increase in the symmetries of nature at a fundamental level. At low energies most of these symmetries must be hidden if string theories are to describe the real world. Ultimately we would like to determine whether or not various symmetries are broken by general topological or dynamical considerations. However at our present level of understanding there are many possible vacua for string theories and many possibilities for which symmetries are broken and which are unbroken. A more modest question that we can attempt to answer is simply whether or not we can find possible vacua that leave an unbroken gauge group which is close to the SU(3)×SU(2)×U(1) group we observe at low energies and with six of the ten spacetime dimensions

---

*Research supported in part by NSF Grant PHY80-19754.

compactified. Since string theories are more highly constrained than field theories we would like to formulate symmetry breaking directly in the string theory without having to resort to its low-energy field theoretic limit. Various aspects of symmetry breaking in string theories have been discussed in refs. 1, 5. Here I will discuss a method of symmetry breaking which was developed in ref. 2 and which has the advantage of being conceptually simple and reasonably tractable from a computational point of view. Since the $E_8 \times E_8$ heterotic string has the best phenomenological prospects I will discuss symmetry breaking exclusively in this theory.

2. TOROIDAL COMPACTIFICATION AND WILSON LINES

A simple but powerful way of breaking symmetries in string theory is to consider strings propagating on a space "modded out" by the action of a discrete group. For example if we want to break the ten-dimensional Lorentz group down to the four-dimensional Lorentz group by compactifying six of the coordinates the simplest consistent possibility is simply to take the six coordinates to parametrize a six-torus. In other words we mod out Euclidean six-space R^6 by the action of a Z^6 discrete group generated by six independent translations to form a six torus $T^6 = R^6/Z^6$. The six translations generate a lattice Λ and requiring that the theory be invariant under translation by a lattice vector requires that the center of mass momentum of the string lie on the dual lattice, Λ^*. In a closed string theory we must also allow for states with non-zero winding number on the torus. Viewed as strings on R^6 these are states that close only up to an element of the Z^6 group of translations.

This simple example has several interesting generalizations. In each case the procedure is to find a discrete symmetry G acting on the string theory, to project onto G-invariant states, and to include "twisted" string states that close only up to transformations by elements of G. From a path integral point of view the projection onto G-invariant states is a sum over all possible τ boundary conditions on the string. Since modular transformations mix σ and τ we must also include the twisted states with all possible σ boundary conditions in order to maintain modular invariance.

In order to describe gauge symmetry breaking we can take G to act also on the gauge degrees of freedom of the heterotic string. For example we can accompany each translation on R^6 in the previous example by a translation on the maximal torus of $E_8 \times E_8$. The group G is still Z^6 but each group element has the form $g=\alpha\beta$ where α acts on R^6 and β acts as a translation by $\pi\delta^I$ on the internal degrees of freedom, $\beta = e^{2\pi p^I \delta^I}$. Here p^I, $I = 1..16$ is the center of mass momentum in the bosonic formulation of $E_8 \times E_8$, the factor of π has been inserted for convenience and δ^I must not a lattice vector if the translation is to act non-trivially. Projecting onto G-invariant states correlates the momentum on T^6 with the internal quantum numbers or "internal momenta," leading in the field theory limit to what we would describe as symmetry breaking by Wilson lines. In string theory we must also include the twisted sectors. In this example these are string states that wind around T^6 and which have boundary conditions $X^I(\pi) = X^I(o) + \pi\delta^I$ for the $E_8 \times E_8$ coordinates. Using the normal mode expansion

$$X^I(\tau+\sigma) = x^I + p^I(\tau+\sigma) + \frac{i}{2} \sum_{n \neq o} \frac{\alpha_n^I}{n} e^{-2in(\tau+\sigma)} \tag{1}$$

we see that in this sector $p^I = \delta^I + L^I$ where L^I is a lattice vector. Since the weight vectors p^I label the U(1) charges of the Cartan subalgebra of $E_8 \times E_8$ we see that in this sector the string states have fractional charges since δ^I is not a lattice vector if we have non-trivial symmetry breaking. In more interesting examples such as Calabi-Yau spaces the fundamental group will not be a product of Z factors but will be finite. For a manifold with fundamental group Z_n we can again introduce Wilson lines by associating translations around the non-contractible loops with group transformations. Now however n windings are equivalent to zero windings and thus $n\delta^I$ must be a lattice vector so the fractional charges are 1/n times the usual charges.[3]

In the case of a non-abelian fundamental group we cannot represent the action on $E_8 \times E_8$ in such a simple way. Probably the simplest possibility is to use the fermionic formulation of $E_8 \times E_8$ which has a linearly realized $O(16)\times O(16)$ symmetry. As long as we can imbed the fundamental group in $O(16)\times O(16)$ we can define its action on the fermions which transform as $(16,1) + (1,16)$. There are certain cases in which we can still use the bosonic formulation however which are of interest since they bear a resemblance to compactification on orbifolds which will be discussed in the next section.

In the bosonic formulation of the heterotic string there are translational symmetries which represent the action of elements in the Cartan subalgebra and also discrete rotational symmetries which correspond to the automorphisms of the $E_8 \times E_8$ lattice. Thus we could also represent certain group actions as rotations on the bosonic coordinates $X^I \to W^{IJ} X^J$ with W^{IJ} being a $O(16)$ rotation which is an automorphism of the E_8 lattice, i.e. an element of the Weyl group of E_8. As a simple example consider a toy model where we represent the action on E_8 by the $O(16)$ rotation $w = \mathrm{diag}\,(-1,-1,-1,\ldots -1)$. In terms of an automorphism of the E_8 Lie algebra w corresponds to the canonical involution $E_\alpha \to E_{-\alpha}$, $H^I \to -H^I$ in the Cartan-Weyl basis. This involution can be extended to the full E_8 affine Lie algebra.

In the untwisted sector the action of w on the previously massless states is given by

$$w|p^I\rangle = |-p^I\rangle \;, \quad w\,\alpha_{-1}^I|0\rangle = -\alpha_{-1}^I|0\rangle \qquad (2)$$

so that the $\underset{\sim}{496}$ adjoint of E_8 splits into 120 states of eigenvalue +1 given by $1/\sqrt{2}\,(|p^I\rangle + |-p^I\rangle)$, $(p^I)^2 = 2$ and 128 states with eigenvalue -1 given by $\alpha_{-1}^I|0\rangle$, $1/\sqrt{2}\,(|p^I\rangle - |-p^I\rangle)$. In terms of Wilson lines the states of eigenvalue +1 can have zero momentum on the six-dimensional compact space and thus remain massless while the states of eigenvalue -1 acquire mass since they must have momenta on the six-dimensional space in order to be invariant. Thus we have broken E_8 down to $O(16)$ with the $\underset{\sim}{120}$ adjoint of $O(16)$ remaining massless and the spinor $\underset{\sim}{128}$ becoming massive.

In the twisted sector we have $X^I(\pi) = -X^I(o)$. Using the normal mode expansion (1) we see that this requires half integer mode numbers $n \varepsilon Z+1/2$, $p^I=0$ and $x^I=-x^I$. Since the center of mass coordinate is only defined modulo a lattice vector, $x^I=-x^I$ has non-zero solutions corresponding to the fixed points of the transformation w. w has 2^8 fixed points on the E_8 lattice but there is a subtlety due to the fact that the X^I have only left-moving modes. View the E_8 lattice Γ_8 as the abelian group generated by translations by the simple roots. Then in the heterotic string one must work not with Γ_8 but with a projective representation, $\bar{\Gamma}_8$. The two-cocycle in the vertex operators which determines this projective representation is necessary in order to ensure the proper commutation relations and is also required in order to have dual tree amplitudes. Thus rather than solving $x^I=-x^I \bmod \Gamma_8$ or equivalently $2x^I \varepsilon \Gamma_8$ we should look for a faithful representation of the group $\overline{\Gamma_8/2\Gamma_8}$. The lowest dimensional faithful representation of this group has dimension $2^4=16$ so in effect we have to take the square root of the number of fixed points. This is also required in order to maintain modular invariance.

The half integer moding in the twisted sector also changes the zero-point energy. This change could be determined say by demanding Lorentz invariance in light-cone gauge but a quicker method is to simply regulate the zero point energy following Brink and Nielsen.[4] A bosonic oscillator with moding $n \varepsilon Z+a$, $o<a<1$, has a zero point energy

$$\sum_{n \varepsilon Z+a} \frac{n}{2} = \sum_{n \varepsilon Z} (n+a)^{-s} \Big|_{s=-1} = -\frac{1}{24} + \frac{1}{4} a(1-a). \qquad (3)$$

Thus the normal ordering term for the left-handed modes (including the 8 untwisted spacetime coordinates) is 16 (-1/24) + 8 (1/48) = -1/2 as compared to -1 in the untwisted sector. Note that the lowest 16-dimensional state and the states built on it are in a different SO(16) conjugacy class from the SO(16) states in the untwisted sector thus giving a non-abelian version of charge fractionalization.

Since w can be chosen to lie in a Cartan subalgebra of E_8 we would expect that twisting by w should be equivalent to a shift by some vector. It is not hard to check that this is indeed the case. In an orthonormal basis e_i, i=1..8 the roots of E_8 are given by

$$\pm e_i \pm e_j \qquad i \neq j$$

$$\frac{1}{2}\left(\pm e_1 \pm e_2 \ldots \pm e_8\right) \quad \text{even number of + signs} \tag{4}$$

and a twist by w is equivalent to a shift by $\delta = e_1$, which is half of a lattice vector. In fact for E_8 all automorphisms are inner, i.e. they can be represented as E_8 group elements so in this context any twist is equivalent to a Wilson line or in other words to accompanying a translation around a non-contractible loop by a gauge transformation. This is not quite true for $E_8 \times E_8$. The Weyl group of E_8 consists of the Weyl group for each E_8 plus an involution that interchanges the two E_8's. Modding out by this involution is not equivalent to a Wilson line and breaks $E_8 \times E_8$ down to E_8. It would be interesting if this could somehow be incorporated into realistic models thus getting rid of the "other" E_8 and its associated problems.

We have mentioned three ways of describing the action of a gauge transformation on the $E_8 \times E_8$ affine Lie algebra of the heterotic string; namely by a translation on the $E_8 \times E_8$ torus for an element in the Cartan subalgebra, by a discrete rotation for elements in the Weyl group, and by an O(16) transformation in the fermionic formulation for $E_8 \times E_8$ elements that lie in O(16)×O(16). While the latter description is certainly the most general the description in terms of twists by automorphisms of the $E_8 \times E_8$ lattice is rather similar to the construction of the Calabi-Yau Z-manifold discussed in Ref. 5. This suggests that we try to treat the compactification from ten to four dimensions also in terms of twisted strings.

3. COMPACTIFICATION ON ORBIFOLDS

Let us first recall the construction of the Z-manifold. One starts with a six torus T^6 which is a direct product of three two-tori, $T^6=T_1\times T_2\times T_3$ with T_i defined by identifying its complex coordinate z_i by $z_i \sim z_i+1$, $z_i \sim z_i + e^{i\pi/3}$. The lattice defining T^6 has various automorphisms, the one we will focus on is given by

$$\alpha: \quad z_i \to e^{2\pi i/3} z_i \tag{5}$$

which generates the cyclic group Z_3. One then considers the space T^6/Z_3 obtained by identifying points under the action of α. Since α has fixed points the resulting space is not a manifold but is what is called an orbifold or a V-manifold. As discussed in [5] it is possible to construct a Ricci-flat Kahler manifold from this orbifold by removing each one of the $3^3=27$ fixed points and gluing back in an appropriate manifold. This results in a Calabi-Yau manifold with Euler number 72 which leads to 36 generations of E_6 after embedding the SU(3) spin connection in one of the E_8 factors.

In analogy with twists acting on E_8 we might wonder whether it is possible to discuss string propagation on the Z-orbifold directly without worrying about repairing the singularities to form a manifold. Perhaps surprisingly the answer is yes. In fact string propagation on the orbifold captures much if not all of the topological structure of the resulting manifold in the cases where it is known how to repair the singularities. Let us review how this works for the simple example of the Z-orbifold discussed in [2].

Following our previous discussion we will have three sectors of string propagation, the untwisted sector, a sector twisted by α, and a sector twisted by $\alpha^2=\alpha^{-1}$. If α acts only on space-time degrees of freedom as in eq. (5) extended appropriately to space-time fermions then the twisted sectors will not contain physical states in general. To see this recall that in a closed string thoery the mass operator must receive equal contributions from left- and right-moving modes or in other words $L_0=\bar{L}_0$. This follows from invariance under rigid shifts in σ since $L_0-\bar{L}_0$ is the generator of infinitesimal shifts of σ. In the twisted sector the normal ordering terms in L_0 and $\bar{L}_0$ are shifted differently since the right movers consist of 8 bosonic and 8 fermionic space-time degrees of freedom while the left-movers consist of 8 bosonic space-time degrees of freedom and either 16 bosonic or 32 fermionic internal degrees of freedom. One way of ensuring that the normal ordering contributions match is to imbed the Z_3 group generated by α in an SU(3) subgroup of E_8 which amounts to a discrete version of imbedding the spin connection in the gauge group.

In analogy to the previous discussion of Wilson lines we consider the Z_3 group generated by elements of the form $\gamma=\alpha\beta$ where β acts on the $E_8 \times E_8$ degrees of freedom. In the bosonic formulation we can take β to act as a translation $\beta=e^{2\pi i p^I \delta^I}$. In terms of the previous basis for the E_8 roots we can take the SU(3) roots to be $\pm(e_1+e_2)$, $\pm(e_1+e_3)$, and $\pm(e_2-e_3)$. An appropriate choice for δ is then

$$\delta = \frac{1}{3}\,(2e_1 + e_2 + e_3). \tag{6}$$

In the untwisted sector the right-moving ground state consists of a vector and spinor of SO(8). The decomposition of SO(8) in terms of SU(3) and the four dimensional helicity SO(2) gives

$$8_v = 3(0) + \bar{3}(0) + 1(1) + 1(-1)$$

$$8_s = 3(-1/2) + \bar{3}(1/2) + 1(1/2) + 1(-1/2) \tag{7}$$

where the helicity is listed in parantheses. Under the action of γ which lies in the center of SU(3) these states have eigenvalues determined by their triality, i.e. $e^{2\pi i/3}$ for the $\underset{\sim}{3}$, $e^{-2\pi i/3}$ for the $\underset{\sim}{\bar{3}}$, and 1 for $\underset{\sim}{1}$. γ invariant states are formed by combining these states with left-moving states with the complex conjugate eigenvalue. The left-moving states consist of the 8_v from the left-moving space-time coordinates and the adjoint of $E_8 \times E_8$ given by

$$\alpha^{I}_{-1}|0\rangle_L \;, \quad |p^I, (p^I)^2 = 2\rangle_L \tag{8}$$

It is not hard to check that these states decompose under $E_6 \times SU(3) \times E_8$ as given below

γ eigenvalue	$E_6 \times SU(3$
1	(1,1,248), (78,1,1), (1,8,1)
$e^{2\pi i/3}$	(27,3)
$e^{-2\pi i/3}$	$(\overline{27}, \overline{3})$

(9)

The γ-invariant states arising from $8_v \times 8_v$ and $8_s \times 8_v$ consist of the supergravity multiplet with helicities (2,3/2,-3/2,-2) and ten super matter multiplets of helicity (1/2,0) and (-1/2,0). The γ-invariant states with gauge quantum numbers consists of the gauge supermultiplet for $E_6 \times SU(3) \times E_8$ and chiral matter supermultiplets with helicity (1/2,0) transforming as (27,3) under $E_6 \times SU(3)$ plus their antiparticles with helicity (-1/2,0) transforming as $(\overline{27},\overline{3})$.

Now consider the sector twisted by γ. If we form complex combinations of the compactified string coordinates, $Z_1 = X_4 + iX_5$, $Z_2 = X_6 + iX_7$, $Z_3 = X_8 + iX_9$, then these coordinates satisfy the boundary conditions

$$Z_i(\pi) = e^{2\pi i/3}\, Z_i(0). \tag{10}$$

Since the Z_i have both left- and right-moving modes this implies that the center of mass coordinate must be at one of the 27 fixed points of α. These twisted states are rather peculiar in that they sit at definite locations and have zero center of mass momentum. In particular this implies that they carry no knowledge of the size of the orbifold. The boundary condition on the Z_i's implies that the normal modes have $Z \pm 1/3$ quanta of oscillation for the compactified coordinates while the two uncompactified coordinates in light-cone gauge have integral quanta. Since γ respects space-time supersymmetry the space-time fermions have the same quanta of oscillation so that the normal ordering term is still zero.

To determine the massless modes in the sector twisted by γ we tensor the left-moving modes with right-moving modes to form γ-invariant states at each one of the 27 fixed points. The right-moving fermions have only two zero modes and quantization of these zero modes leads to two states with helicity (0,1/2) and γ eigenvalue 1. For the left-movers we have twisted boundary conditions for the spacetime bosons and also for the $E_8 \times E_8$ bosons. The latter satisfy

$$X^I(\pi) = X^I(o) + \delta^I \tag{11}$$

with δ^I given by Eq. (6). The center of mass momenta therefore lie on a "shifted" lattice $p^I = L^I + \delta^I$, $L^I \in \Gamma_8 \times \Gamma_8$. The twist (10) on the spacetime bosons implies a normal ordering term in the left-hand sector which is -2/3 using the formula (3). Massless states must therefore satisfy $N + \frac{1}{2}\Sigma(p^I)^2 - \frac{2}{3} = 0$, and will have eigenvalue $e^{\pi i p^I p^I}$ times the eigenvalue of the left-handed vacuum under the action of γ. It is easy to check that there are 27 states corresponding to $p = \delta + L$ with roots L having components along the e_i of the form

$$\begin{array}{ll} (-1,0,0,0..\pm1..0) & 10 \\ (0,-1,-1,(0)^5) & 1 \\ 1/2(-1,-1,-1,(\pm1)^5) & 16 \end{array} \tag{12}$$

corresponding to the decomposition of the $\underset{\sim}{27}$ of E_8 into SO(10) multiplets. One can use the chiral anomaly or the requirement of modular invariance to check that these states all have eigenvalue 1 under γ. The sector twisted by γ thus gives twenty seven (one for each fixed point) chiral matter multiplets in the $\underset{\sim}{27}$ of E_6. There are also nine E_6 singlet states with eigenvalue 1 arising from states created by

the fractionally moded space-time coordinates of the form $\alpha^{i}_{-1/3}$ $\alpha^{j}_{-1/3}|o\rangle_L$, $\bar{\alpha}^{i}_{-2/3}|o\rangle_L$, $i,j = 1,2,3$.
The sector twisted by $\gamma^2=\gamma^{-1}$ yields the antiparticles of these so altogether we find 36 chiral $\underset{\sim}{27}$'s of E_6.

This agrees with what one obtains by studying the manifold formed by repairing the singularities in the field theoretic limit. In that case the number of generations is one half of the Euler characteristic of the manifold. The Euler characteristic of the Z-manifold is easily calculated.[5] One starts with the six-torus T^6 with Euler characteristic zero and removes the 27 fixed points to form a space $\tilde{T}^6$ with Euler characteristic -27. Identifying points under the motion of the Z_3 group generated by α gives $\tilde{T}^6/Z_3$ with Euler characteristic -9. The Z-manifold is obtained by gluing in at the 27 fixed points a space X with Euler characteristic 3 to yield a total Euler characteristic of -9+3.27=72. In fact for the Z-manifold each generation is associated with a non-contractible four surface[6] with 9 of the four surfaces being those that come from T^6 that are invariant under the Z_3 symmetry and 27 coming from the fact that each X manifold used to repair the singularities has a non-contractible four surface. In the orbifold limit the size of the X manifold shrinks to zero and the wave functions of 27 of the chiral multiplets become concentrated on the fixed points of the Z_3 transformation corresponding to the twisted sectors of string propagation.

This analysis can easily be extended to more general models by starting with different six tori and modding out by more complicated crystallographic subgroups of SU(3). The various models that can be obtained this way are discussed in the second reference in [2]. One advantage of orbifolds is that they should make possible explicit calculations of Yukawa couplings as well as string loop effects. It is also possible that symmetry breaking effects may tend to drive Calabi-Yau spaces to the singular orbifold limit although this is just conjecture at present.

4. HETEROTIC MOONSHINE

One disappointing aspect of the previous discussion is the enormous choice of twistings or symmetry breakings that appear to lead to consistent string theories. We do not yet understand the principles that determine the correct (and hopefully unique) vacuum of string theory. There is a remarkable construction of the monster group F_1 in string theory where a particular twist is picked out from a mathematical point of view. This construction is reviewed in the talk by J. Lepowsky at this workshop. Although there is no clear connection between the monster and physics the elements of the construction of the monster in ref. 7 are so analogous to attempts to describe symmetry breaking in string theory that one feels that some elements of the construction may provide a clue as to the correct choice of twistings in realistic string theories.

The construction of the monster in ref. 7 can be naturally embedded in certain two-dimensional heterotic string theories. In generalizing the heterotic string one possibility is to take the left-moving modes to be those of the 26-dimensional bosonic string compactified on a 24-dimensional even self-dual lattice and the right-movers to be those of the two-dimensional N=2 superstring. However the N=2 theory has no transverse modes and is actually a field theory, not a string theory since there are no physical massive modes.

A more promising approach[8] is to again take the right-movers to be those of the ten-dimensional superstring but compactified on an eight-dimensional, even, self-dual lattice, i.e. Γ_8. There are 24 even self-dual lattices in 24 dimensions so there are 24 possible two-dimensional heterotic string theories of this type corresponding to which lattice is chosen for the left-moving modes. The most interesting of these is the one corresponding to the Leech lattice, Λ, which is the unique 24-dimensional even self-dual lattice without points of $(\text{length})^2=2$. In light cone gauge the physical degrees of freedom in this theory have the normal mode expansions

$$X^I(\tau+\sigma) = x^I + \tilde{p}^I(\tau+\sigma) + \frac{i}{2}\sum_{n\neq 0}\frac{\tilde{\alpha}_n^I}{n}e^{-2in(\tau+\sigma)} \qquad I=1..24, \quad \tilde{p}\epsilon\Lambda$$

$$X^i(\tau-\sigma) = x^i + p^i(\tau-\sigma) + \frac{i}{2}\sum_{n\neq 0}\frac{\tilde{\alpha}_n^i}{n}e^{-2in(\tau-\sigma)} \qquad i=1..8, \quad p\epsilon\Gamma_8 \qquad (13)$$

$$S^a(\tau-\sigma) = \sum_n S_n^a e^{-2in(\tau-\sigma)} \qquad a=1..8$$

The mass operator is

$$\frac{1}{4}(\text{mass})^2 = N + \tilde{N} - 1 + \frac{1}{2}\sum_I(\tilde{p}^I)^2 + \frac{1}{2}\sum_i(p^i)^2 \qquad (14)$$

and invariance under shifts in σ gives the constraint

$$N + \frac{1}{2}\sum_i(p^i)^2 = \tilde{N} - 1 + \frac{1}{2}\sum_I(p^I)^2 \quad . \qquad (15)$$

The right-handed ground state is again 16 dimensional and forms a representation of the S^a zero mode algebra. The massless states are formed by tensoring these states with left-moving states with $\tilde{N}=1$ or $\Sigma_I(p^I) = 2$. For the Leech lattice we thus get 24 massless supermultiplets of the form $|i \text{ or } a\rangle_R \times \tilde{\alpha}_{-1}^I |o\rangle_L$. At the next mass level for the left-moving states there are 24 states of the form $\tilde{\alpha}_{-2}^I |o\rangle_L$, 12·25 of the form $\tilde{\alpha}_{-1}^I \tilde{\alpha}_{-1}^I |o\rangle_L$ and 196,560 of the form $|p^I, (p^I)^2 = 4\rangle$. In general the number of states at each mass level is given by the partition function which for the left-movers is

$$\frac{\Lambda(q)}{q\prod_n(1-q^n)^{24}} \equiv \frac{\theta_\Lambda(q)}{(\eta(q))^{24}} = q^{-1} + 24 + 196884q + \ldots \qquad (16)$$

where $\theta_\Lambda(q) = \sum_{\ell\epsilon\Lambda} q^{\ell^2/2}$ is the theta function for the Leech lattice and $\eta(q)$ is Dedekind's η-function. Expect for the constant term this partition function is equal to the modular function $j(q) = q^{-1} + 744 + 196884q + \ldots.$

There are many mathematical connections between the monster group and modular functions which go under the general name of "monstrous moonshine." They started with the observation that each coefficient in j(q) except for the constant term is a sum of dimensions of irreducible representations of the monster! In terms of string theory this means that at each mass level the states fall into monster multiplets except for the massless states. The question is then clearly whether one can actually construct the monster group as a symmetry group of the states of the string theory. For this to be possible the massless states must be removed but this can be done by using a twist that is the analog of the canonical involution discussed earlier in the context of breaking E_8 down to O(16). We thus consider the involution θ with $\theta\,\tilde{\alpha}^I_{-1}\,|o\rangle_L = -\tilde{\alpha}^I_{-1}|o\rangle_L$, $\theta|p^I\rangle_L = |-p\rangle_L$ etc. The projection onto θ invariant states removes the 24 massless states and the states at the next mass level of the form $\tilde{\alpha}^I_{-2}\,|o\rangle$, $1/\sqrt{2}\,(|p^I\rangle - |-p^I\rangle)$. However to preserve modular invariance the twisted sector must be added with boundary condition $X^I(\pi) = -X^I(o)$ and one must deal with a representation of the group $\overline{\Lambda/2\Lambda}$. The result of this is that the partition function for θ invariant states including untwisted and twisted sectors is as before but with the 24 removed. Furthermore Frenkel, Lepowsky and Meurman have shown that the monster group is in fact realized as a symmetry group of the theory. This is highly non-trivial since elements of F_1 must act between Hilbert spaces with different boundary conditions much as the supersymmetry operator does in the old superstring formolism.

There also appears to be moonshine associated with the superstring part of this construction.[7] The right-moving superstring partition function in this model is

$$\frac{16\theta_{\Gamma_8}(q)\,\prod_n (1+q^n)^8}{\prod_n (1-q^n)^8} = 2(8 + 2048q + 49152\,q^2 + \ldots) \qquad (17)$$

and the coefficients of the terms in parentheses are sums of irreducible representations of another sporadic group, .1, which is associated with automorphisms of the Leech lattice.

The connection between sporadic groups, modular functions and string theory is mysterious and probably quite deep. It may also have nothing to do with the real world. However it is a hint that there are very subtle mathematical structures in the heterotic string which we will probably have to understand better if we are to determine whether or not string theories describe the real world.

Acknowledgement

I would like to thank L. Dixon and C. Vafa for explanations of how one determines the charges of twisted string states and J. Lepowsky for discussions.

REFERENCES

1. E. Witten, Nucl. Phys. B258 (1985) 75;
J.D. Breit, B.A. Ovrut and G. Segrè, Phys. Lett. B158 (1985) 33;
A. Sen, Phys. Rev. Lett. 55 (1985) 33.
2. L. Dixon, J. Harvey, C. Vafa and E. Witten, Strings on Orbifolds I and II, Princeton preprints (1985).
3. X.-G. Wen and E. Witten, Electric and Magnetic Charges in Superstring Models, Princeton preprint (1985).
4. L. Brink and H.B. Nielsen, Phys. Lett. 45B (1973) 332.
5. P. Candelas, G. Horowitz, A. Strominger and E. Witten, Nucl. Phys. B258 (1985) 246.
6. A. Strominger and E. Witten, to appear in Comm. Math. Phys.;
A. Strominger, preprint NSF-ITP-85-109.
7. I.B. Frenkel, J. Lepowsky and A. Meurman, in "Vertex Operators in Mathematics and Physics, Publications of the Mathematical Sciences Research Institute No. 3 (Springer, Berlin, 1984).
8. J. Harvey and J. Minahan, in preparation.

Natural Extensions of String Theories

Freydoon Mansouri

Physics Department, University of Cincinnati,
Cincinnati, Ohio 45221

Abstract

A method of constructing dual amplitudes corresponding to objects with more than one spacial dimension is described. The 4-point function is explicitly evaluated and shown to be of closed string type. Factorization is carried out and the ground state vertex operator is explicitly given. The possibility of deriving such models from an action principle is discussed.

I. Introduction

String theories[1] provide us with a rich mathematical structure which may pave the way for a consistent unified theory of all interactions. Among the many intriging features of these theories, there are some which are still not well understood. One of these is the critical dimension associated with each string theory. Both unitarity and Lorentz invariance point to critical dimensions 26 or 10. Then to reconcile these theories with successful 4-dimensional theories such as electroweak, QCD, and GUTs, one must subject the higher dimensional string theories to some sort of dimensional reduction. That is, one must regard the extra dimensions as compactified or trade them with internal symmetries. In this sense, one may consider the string theories as concrete realizations of the Kaluza-Klein approach. In fact, one of the modern revivals of the Kaluza-Klein approach was motivated by this very fact.[2]

On the other hand, recent work on string theories in 10 dimensions has revealed another intriguing feature of these theories: To be anomoly free, they must be endowed with either O(32) or $E_8 \times E_8$ internal symmetries. So, one must deal with strings with internal symmetries. This appears to run counter to the spirit of Kaluza-Klein because now we have <u>both</u> internal symmetry and the extra dimensions. It is therefore not totally absurd to ask: What is a string with internal symmetry?

As a first step in providing an answer to this question, we note that in string theories one assumes that matter at very high energies is organized into one dimensional objects and asks what critical space-time dimensions can support a consistent theory of such objects. In this case, matter distribution is already "compactified". Further compactification can only be applied to the space-time itself.

In the work that I will describe below, I will take the point of view that a string with internal symmetry is not a primary form of matter but, in the spirit of standard Kaluza-Klein approach, the compactified left over of a form of matter which was initially extended in more than one dimension. In this case, the compactification may occur either with respect to the matter form or the space-time, or both.

To study the extent to which the higher dimensional matter configurations have acceptable physical properties, it will be helpful, for comparison purposes, to summarize the fundamental properties which make the string models so special. Using this as a guide, I will present a method of constructing dual n-point amplitudes for the extended objects, which parallels the development of string theories. I will explicitly evaluate the 4-point function and show that it is of closed string type. These models are then factorized and the corresponding ground state vertex operators are constructed. Finally, I will discuss the prospects of deriving such models from an action.

II. Why are strings so special?

It will be recalled that string models originally grew out of Regge pole phenomenology and S-matrix theory.[1] The first major departure from the standard Regge pole theory was the formulation of a 4-point function by Veneziuno, which incorporated Regge assymptotic behavior as well as duality and crosssing symmetry. The n-point generalization of this model was soon given by several authors. Among these n-point representations, I mention two which will be used as models for writing down amplitudes associated with objects with more than one spacial dimension. The first was the representation of Koba and Nielsen. It was here that the concepts of symmetry and invariance first made their appearance in dual models. The SU(1,1) invariance which emerged in this representation can be used as an alternative basis for the construction of dual amplitudes.

The second representation of dual n-point amplitudes which I will make use of in the sequel was the one given by Nambu in terms of string oscillators[4]. After factorizing the n-point functions in terms of oscillators he suggested a picture of hadrons as massless strings each represented by the string variable $Y^{\mu}(\tau,\sigma)$ which satisfies the wave equation

$$\left[\frac{\partial^2}{\partial\tau^2} - \frac{\partial^2}{\partial\sigma^2}\right] Y^{\mu}(\tau,\sigma) = 0 \qquad \text{(II.1)}$$

the corresponding Lagragian is

$$\mathcal{L}_1 = \frac{1}{2}\left[\left(\frac{\partial Y^{\mu}}{\partial\tau}\right)^2 - \left(\frac{\partial Y^{\mu}}{\partial\sigma}\right)^2\right] \qquad \text{(II.2)}$$

The hadronic interactions would then be described as breaking and joining of the strings, where the vertex operator for the ground state was given by

$$: e^{ik.Y} : \qquad \text{(II.3)}$$

A dual theory formulated along these lines was a major improvement over the previous c-number formulations. For a number of reasons, however, it did not gain wide acceptance. One reason was that in its present form this theory has time-like oscillations if it is to be Lorentz invariant. But then in the absence of any built mechanism to eliminate them, this leads to ghosts and the violation of unitarity. Another reason had to do with the role of the variables τ and σ and their choice on the world sheet of the string. Clearly, the Lagrangian (II.2) is not related to any intrinsic property of the world sheet of the string.

To deal with these and other problems, a number of improvements were suggested. To remove the time-like oscillations from the spectrum, it was necessary to impose a set of subsidiary conditions on the states, one for each oscillator mode. Such a set was invented by Virasoro[5] and constitutes the Virasoro algebra. Although _ad hoc_, these conditions were later shown to reduce the Hilbert space of the physical states to the transverse ones. In another development, it was pointed out by Nambu[6], and later by Chang and myself[7] and by Goto[8], that if one proceeds in pure analogy with point particle mechanics, then the most natural actions for the string would be those related to the geometry of the world sheet of the string. Among the Lagrangians suggested were[6]

$$\mathcal{L}_2 = \sqrt{-g} \qquad \text{(II.4)}$$

$$\mathcal{L}_3 = \sqrt{-g}R \tag{II.5}$$

where g = determinent of the metric tensor, g_{ab}, of the world sheet. As pointed out by Nambu, both of these actions were highly non-linear and had no apparent relation to the theory of dual resonance models.

As of late 1970, the string theory of the dual resonance models was faced with a dilemma: On the one hand, there existed a theory based on the Lagrangian $\mathcal{L}_1$, the vertex (II.3), and Virasoro's conditions. The logical consistency of this formalism, its uniqueness, and its physical foundation, all left something to be desired. On the other hand, there were geometrical possibilities given by $\mathcal{L}_2$ and $\mathcal{L}_3$, which were at the time purely objects of curiosity with no known connection to dual resonance models.

The modern string theory came into being when it was pointed out by Chang and me[7,9] that the spectrum and the ghost eliminating conditions of the dual resonance theory can both be derived from the Lagrangian $\mathcal{L}_2$. Making use of the properties of two dimensional manifolds, it was shown[7] that by an appropriate coordinate transformation it is always possible to write the line element

$$ds^2 = g_{ab}d\xi^a d\xi^b$$

in the forms

$$\begin{aligned} ds^2 &= g_{+-}du^+du^- = g_{00}(d\tau^2 - d\sigma^2) \\ u^{\pm} &= \frac{1}{\sqrt{2}}(\tau \pm \sigma) \end{aligned} \tag{II.6}$$

One immediate consequence of this result is that by the choice of (conformal) coordinates (II.6) with respect to which

$$\begin{aligned} g_{00} + g_{11} &= 0 \\ g_{01} = g_{10} &= 0 \end{aligned} \tag{II.7}$$

the Lagrangian $\mathcal{L}_2$ reduces to $\mathcal{L}_1$. thus, we have the equivalence

$$\mathcal{L}_2 \rightleftarrows \left\{ \mathcal{L}_1 \quad \text{and} \quad \begin{matrix} \dot{x}^2 + x'^2 = 0 \\ \dot{x} \cdot x' = 0 \end{matrix} \right\} \tag{II.8}$$

The constraints (II.7) or (II.8) are the geometrical basis for the relevance of the Virasoro algebra. After quantization, they become operator conditions on states, and their Fourier components give the elements of the Virasoro algebra. This is how such by now familiar notions as "reparametrization invariance," "conformal gauge," and the identification of the Virasoro algebra as generators of the co-ordinate transformation on the string world sheet originally appeared in string theory. It was also shown in these papers that the ground state of the full spectrum would have to be a tachyon.

The necessity for explicitly checking the Lorentz invariance of a ghost-free string theory was first brought up in the paper by Chang and meself[7]. But we left the question unanswered. About a year later, it was taken up by GGRT[10] who, starting from our conformal gauge, further fixed the gauge to be the light cone gauge. Then by constructing the Lorentz generators in this gauge, they showed that the string theory is Lorentz invariant only if the dimension, D, of space-time is equal to 26. This was consistent with an earlier result by Lovelace[11], in which the need for D=26 arose from the consistency of one-loop corrections.

Faced with the prospect of having to abandon the string theory as a possible physical theory, a number of attempts were made to establish contact with four space-time dimensions. By this time, there was sufficient evidence to conclude that the string theory could not directly describe hadrons. So it was natural to see if they could be interpreted as constitutents of hardrons. In that case, it would no longer be necessary to quantize the theory according to the ordinary statistics. Following this line of argument, string models were quantized according to parastatistics[12]. It was found that the corresponding quantum theories have critical dimensions which depend on the order of parastatistics. They are given by

$$D = 2 + \frac{24}{q} \qquad \text{(Bosonic string)}$$

$$D = 2 + \frac{8}{q} \qquad \text{(Superstrings)}$$

Here q is the order of parastatistics. Since there is a correspondence between parastatistics and exact internal symmetries, the number q may also be interpreted as the index of an SO(q) symmetry. This is how internal symmetry as a symmetry continuously distributed over the string was first introduced in string theory.

From the brief discussion presented above, it is clear why strings are so special: It is because their world histories are two dimensional surfaces, and two dimensional manifolds have special properties not shared by higher dimensional manifolds. In particular, the transformation

$$g_{ab}d\xi^a d\xi^b \rightarrow g_{00}\ (d\tau^2 - d\sigma^2)$$

is a geometrical transformation, so that the imposition of the Virasoro constraints is not ad hoc. As a result, a string theory with or without internal symmetry is mathematically more tractable than a theory of a higher dimensional object.

III. N-Point Functions for Higher Dimensional Objects[13]

In constructing n-point dual amplitudes for higher dimensional objects, the most natural point of departure from the string theories appears to be the Koba-Nielsen representation metnioned above[14]. It will recalled that this representation consists of SU(1,1) invariant amplitudes

$$\begin{aligned} B_N &= \int \prod_{i=1}^{N} dz_i \left[\frac{dz_a dz_b dz_c}{(Z_a - Z_b)(Z_b - Z_c)(Z_c - Z_a)}\right]^{-1} \\ &\quad \times \prod_{1 \leq i < j \leq N} |Z_i - Z_j|^{-2P_i \cdot P_j - C_{j-i+1}} \\ &\equiv \int d^2v\ K(Z_1, \ldots, Z_n) \end{aligned} \qquad \text{(III.1)}$$

where a,b,c refer to any three points on the unit circle in the complex plane, along which the integrations are carried out in a specified manner. This fixing of the three points and the division by the corresponding constant expression can be carried out because of the SU(1,1) invariance of B_N. The N-point dual amplitude is then given by

$$A_N = \sum_{\substack{\text{inequiv.} \\ \text{perm.}}} B_N(P_1, \ldots, P_N) \qquad \text{(III.2)}$$

The complex variables Z, may be identified as the wick rotated combinations of the string variables τ and σ. That is, $z = x+iy$. Therefore, the SU(1,1) invariance of the amplitudes A_N has the same origin as the global (conformal) symmetry of the two dimensional manifolds. This correspondence provides the clue for writing down N-point dual amplitudes for objects with more than one spacial extension. Consider, for example, a three dimensional object. It's time evolution sweeps out a four dimensional surface in a D-dimensional space-time. The conformal group for the 4-dimensional world surface is SO(4,2) ~ SU(2,2). We must then look for dual amplitudes with global SO(4,2) symmetry. In simplest models of this type, we write the N-point amplitude in the form

$$A_N = \int d^4vK(x_1^\mu, x_2^\mu, \ldots, x_N^\mu) \qquad \text{(III.3)}$$

and require that the kernel K and the volume element d^4V be SO(4,2) invariant. Here x_j^μ, $\mu = 0,1,2,3, j=1,\ldots,N$, play the same roles as the string variables τ and σ.

Arbitrary conformal transformations are composed of

$$\begin{aligned} &\text{Translations:} && x^\mu \to x^\mu + a^\mu \\ &\text{Lorentz rotations:} && x^\mu \to \Lambda^\mu{}_\nu x^\nu \\ &\text{scale transformations:} && x^\mu \to \lambda x^\mu \\ &\text{inversions:} && x^\mu \to -\frac{x^\mu}{x^2} \end{aligned} \qquad \text{(III.4)}$$

The invariance under these transformations of the kernel K shows that it can only be a function of the invariant cross (anharmonic) ratios:

$$K(x_1,\ldots,x_N) = K(R_1(x), R_2(x), \ldots) \qquad \text{(III.5)}$$

For example, a 4-point function depends on the two independent cross ratios

$$R_1 = \frac{(x_1-x_2)^2(x_3-x_4)^2}{(x_1-x_3)^2(x_2-x_4)^2} \; ; R_2(x) = \frac{(x_1-x_4)^2(x_2-x_3{}^2)}{(x_1-x_3{}^2)(x_2-x_4)^2} \qquad \text{(III.6)}$$

Similarly, one can check explicitly that the conformally invariant volume element is given by[15]

$$d^4v = \prod_{i=1}^{N} d^4x_i\left[\prod_{j=1}^{N}(x_j-x_{j+1})^2\right]^{-2} \quad ; \; x_{N+1} = x_1 \tag{III.7}$$

So, just as in Koba-Nielsen representation, the dual tree amplitude for the 3-dimensional object can be written down explicitly for any N. For example, the 4-point function is given by

$$A_4 = \int \prod_{i=1}^{4} d^4x_i \left[\frac{(x_1-x_2)^2(x_3-x_4)^2}{(x_1-x_3)^2(x_2-x_4)^2}\right]^{-d_1-2}\left[\frac{(x_1-x_4)^2(x_2-x_3)^2}{(x_1-x_3)^2(x_2-x_4)^2}\right]^{-d_2-2} \tag{III.8}$$

where d_1 and d_2 are invariant functions of the external momenta. As in the Koba-Nielsen representation, d_1 and d_2 are not fixed by SO(4,2) invariance and, at this stage, they are additional inputs to the formalism.

To evaluate A_4 explicitly, one can make use of the invariance of cross ratios under conformal transformations to fix 3 of the four x_k^μ's and separate a divergent factor:

$$x_1^\mu \to I^\mu \; ; \; x_2^\mu \to \infty \; , \; x_3^\mu \to 0 \tag{III.9}$$

where I^μ is a unit 4-vector. Then the expression for A_4 reduces to

$$A_4 = \int d^4x (x^2)^{-d_1-2}[(I^\mu - x^\mu)^2]^{-d_2-2} \tag{III.10}$$

this expression is to be compared with the 4-point Veneziano amplitude

$$B_4 = \int dx|x|^{-\alpha(s)-1}|1-x|^{-\alpha(t)-1}$$

After regularizing the expression (III.10), one obtains

$$A_4 = \frac{\Gamma(1-d(s))\Gamma(1-d(t))\Gamma(1-d(u))}{\Gamma(1+d(s))\Gamma(1+d(t))\Gamma(1+d(u))} \tag{III.11}$$

where

$$d_1 = d(s)-1 \quad ; \; d_2 = d(t)-1$$

and

$$d(s) + d(t) + d(u) = 1$$

This expression is clearly of the symmetric closed string type. Had we not known where it came from, it would have been difficult to find out whether it corresponds to a one dimensional object or a 3-dimensional one.

Before proceeding to factorization, some comments are in order. It is clear from the derivation of the N-point amplitudes that their structure does not depend on three spacial dimensions. In particular, they hold for any number of odd spacial dimensions where the corresponding conformal group is SO(2N,2). Moreover, the structure of these N-point functions is the same as the Green's functions for the conformally invariant field theories. There is some preliminary indication that string, membrane, etc., theories might be non-trivial solutions of standard conformally invariant quantum field theories[16].

IV. Factorization of the 4-Point Function

The factorization of the 4-point function (III.10) can be carried out by a generalization of the method used by Nambu[4] to factorize the Veneziano model. It will be recalled that in that case the factorization amounted to replacing any one of the exponancials of the form

$$\exp[\frac{\alpha'}{n} k_1 \cdot k_2 x^n] \tag{IV.1}$$

by a vacuum expectation value of the products of two exponancial operators, one depending on k_1 and one depending on k_2. To make use of this approach, we first write A_4 in the form

$$A_4 = \int d\Omega dr (r^2)^{-d} [1-2\text{ch}\,\gamma r+r^2]^{\alpha' k_1 \cdot k_2} \tag{IV.2}$$

where $d\Omega$ contains the angular part of the volume element, and $\mathrm{Ch}\gamma$ is the angle between two 4-vectors. So, in this case factorization must not only split $k_1 \cdot k_2$ but also $\mathrm{Ch}\gamma$. Write

$$[1-2\mathrm{ch}\gamma r+r^2]^{\alpha' k_1 \cdot k_2} = \prod_{m=1}^{\infty} \exp[2\alpha' k_1 \cdot k_2 \frac{\mathrm{chm}\gamma}{m} r^m] \tag{IV.3}$$

Now a unit 4-vector (or more generally an n-vector) can be parametrized as follows:

$$I = (\mathrm{Ch}\beta,\ \mathrm{Sh}\beta\mathrm{Sin}\theta\cos\phi,\ \mathrm{Sh}\beta\mathrm{Sin}\theta\mathrm{Sin}\phi,\ \mathrm{Sh}\beta\cos\theta) \tag{IV.4}$$

Then we can write

$$\mathrm{Ch}\gamma = I_1 \cdot I_2 = \mathrm{Ch}\beta_1 \mathrm{Ch}\beta_2 - \mathrm{Sh}\beta_1 \mathrm{Sh}\beta_2 \cos\theta \tag{IV.5}$$

where

$$\cos\theta = \cos\theta_1 \cos\theta_2 + \mathrm{Sin}\theta_1 \mathrm{Sin}\theta_2 \cos(\phi_1 - \phi_2) \tag{IV.6}$$

From (IV.3) it is seen that if we are to use the factorization method outlined above, we must not only split $\mathrm{ch}\gamma$ but also $\mathrm{Chm}\gamma$. This can be accomplished by noting that

$$2\,\frac{\mathrm{Chm}\gamma}{m} = \lim_{\nu\to 0} \Gamma(\nu)\, C_m^{\nu}(\mathrm{Ch}\gamma) \equiv C_m^{o}(\mathrm{Ch}\gamma) \tag{IV.7}$$

where $C_m^{\nu}(\mathrm{ch}\gamma)$ is the Gegenbauer function of order ν. This allows us to use an addition theorem for Gegenbauer functions to factorize $2\mathrm{Chm}\gamma$. To this end, without going into mathematical details, let us define the functions

$$D^m_{\ell\lambda}(\beta;\theta,\phi) = [K^m_{\ell\lambda}]^{1/2} \mathrm{Sh}^{\ell}_{\beta} C^{\ell}_{m-\ell}(\mathrm{Ch}\beta)\ \mathrm{Sin}\theta^{\lambda}\ C^{\lambda-1/2}_{\ell-\lambda}(\cos\theta)\, e^{i\lambda\phi} \tag{IV.8}$$

Then by a straight-forward computation it can be verified that

$$\frac{2\mathrm{Chm}\gamma}{m} = \sum_{\ell=0}^{m} \sum_{\lambda=0}^{\ell} [D^m_{\ell\lambda}(\beta_1,\theta_1,\phi_1) D^m_{\ell\lambda}(\beta_2,\theta_2,\phi_2) + \mathrm{C.C.}] \tag{IV.9}$$

where C.C. stands for the complex conjugate term. From the structure (IV.8) and (IV.9) we can determine what the analog of the string variable should be in this case. Consider the oscillators $a^{\mu}_{m\ell\lambda}$, $b^{\mu}_{m\ell\lambda}$ and their adjoints, satisfying the commutation relations

$$[a^{\mu}_{m_1\ell_1\lambda_1}, a^{\nu\dagger}_{m_2\ell_2\lambda_2}] = g^{\mu\nu}\delta_{m_1m_2}\delta_{\ell_1\ell_2}\delta_{\lambda_1\lambda_2}$$
$$[b^{\mu}_{m_1\ell_1\lambda_1}, b^{\nu\dagger}_{m_2\ell_2\lambda_2}] = g^{\mu\nu}\delta_{m_1m_2}\delta_{\ell_1\ell_2}\delta_{\lambda_1\lambda_2} \tag{IV.10}$$
$$[a,b] = 0$$

The dynamical variable representing the coordinates of the extended object must then be of the form

$$Y^{\mu}(\beta,\theta,\phi,r) = \sum_{m,\ell,\lambda} [D^{m}_{\ell\lambda}(\beta,\theta,\phi)a^{\mu}_{m\ell\lambda}r^{m} + D^{m*}_{\ell\lambda}(\beta,\theta,\phi)b^{\mu\dagger}_{m\ell\lambda}r^{-m}] \tag{IV.11}$$

From this, we find that the ground state vertex operator for this model is given by

$$V(k,\beta,\theta,\phi,R) = :e^{i\sqrt{2}k\cdot(Y+Y^*)}: \tag{IV.12}$$

where, as usual, dots indicate normal ordering.

It is now straight-forward to verify that in terms of this vertex operator our 4-point function can be written in the form

$$A_4 = \int d^4x\,(x^2)^{d(s)} M_4 \tag{IV.13}$$

where

$$M_4 = \langle k_4 0|V(k_3,x_3)V(k_2,x_2)|0k_1\rangle \tag{IV.14}$$

One can, in fact, go further and write A_4, or generally any A_N in terms of vacuum expectation values of products of vertices and propagators. SO, at least up to this stage, the behavior of the dual tree amplitudes for objects with spacial extensions in more than one dimension parallels those which follow from string theories.

V. Concluding Remarks

One of the initial motivations for the study of dual amplitudes corresponding to objects with more spacial dimensions than one was to search for possible principles which uniquely determine the dimension of matter at very high energies. To see if any such principle exists, clearly one must start with no a priori restrictions on spacial dimensions. We have seen that the dual tree amplitudes obtained without such restrictions have enough desirable features which warrant further studies along these lines.

An interesting question which arises in connection with these amplitudes is whether they have an underlying action principle. One way to deal with this question is to proceed in complete analogy with developments in string theories. Then one would be led to consider geometrical actions which depend on the determinent of the metric tenror, on scalar curvature, or on both. For a 3-dimensionally extended object, consider, e.g., the Lagrangian

$$\mathcal{L} = (-g)^{K} \quad ; \quad g = \det\,(g_{ab}) \qquad a,b = 0,1,2,3 \tag{V.1}$$

where K is an appropriate rational number. As pointed out in section II, this non-linear Lagrangain cannot be reduced to a linear one by mere coordinate transformations unless the world hypersurface is 2-dimensional. So for the case at hand the coordinate conditions (II.7) from which the Virasoro algebra can be derived are no longer a consequence of the parametrization invariance of the theory. One may nevertheless ask if there are conditions of the form (II.7), which reduce the Lagrangian (V.1) to a linear one. For this to be true, we must have conditions which allow the reduction[9]

$$g_{ab}d\sigma^{a}d\sigma^{b} \to g_{00}[(d\sigma^{o})^{2}-(d\sigma^{1})^{2}-(d\sigma^{2})^{2}-(d\sigma^{3})^{2}] \tag{V.2}$$

These are

$$g_{00} = -g_{11} = -g_{22} = -g_{33}$$
$$g_{ab} = 0 \quad , a \neq b \qquad (V.3)$$

These conditions include but are, in general, larger in number than those given in (II.7), so that one must study the requirements under which they are mutually consistent and yet not overcomplete so as to render theory trivial.

Let us consider the possibility that some of the conditions of the type (V.3) can be satisfied automatically. This will happen if the local structure of the 4-dimensional world hypersurface is of direct product type. For example, if two of the spacial dimensions are compactified or if the dependence on them becomes trivial because of the presence of isometries, then the 4-dimensional world hypersurface would, effectively, reduce to a 2-dimensional world sheet with a continuous distribution of internal symmetry. Note that in this case compactification is applied not to the space-time but to the world hypersurface of the extended object. The space-time of such a theory may still have a critical dimension associated with it[17], and the dimensional reduction in the usual sense may have to be applied to it. We also note that an approach to dimensional reduction which is tailor made for application to world hypersurfaces already exists in the literature[18].

The hospitality of Lewes Center for Physics, where this manuscript was written up, is gratefully acknowledged. This work was supported, in part, by the Department of Energy under the contract number: DE-AS-2-76ER02978.

References

1. For reviews see J. Scherk, Rev. Mod. Phys. 47, 123 (1975) and S. Mandelstan, Phys. Rep. 13C, 259 (1974).
2. L. N. Chang, K. Macrae, F. Mansouri, Phys. Lett. B57, 59(1975) and Phys. Rev. D13, 235(1976).
3. M. B. Green and J. H. Schwarz, Phys. Lett. 149B, 117 (1984), D. J. Gross, J. A. Harvey, E. Martinec and R. Rohm, Phys. Rev. Lett. 54, 502 (1985).
4. Y. Nambu, Proc. Int. Conf. on Symmetries and Quark Models, Wayn State Univ., 1969, p. 269, Gordon and Breach 1970.
5. M. A. Virasoro, Phys. Rev. D1, 2933 (1970).
6. Y. Nambu, Lecutres Written for the Copenhagen Symposium, 1970, unpublished.
7. L. N. Chang and F. Mansouri, Phys. Rev. D5, 2535 (1972).
8. T. Goto, Prog. Theor. Phys. 46, 1560(1971).
9. F. Mansouri and Y. Nambu Phys. Lett. 39B, 375(1972).
10. P. Goddard, J. Goldstone, C. Rebbi and C. Thorn, Nucl. Phys. B56, 109(1973).
11. C. Lovelace, Phys. Lett. 32B, 703(1970).
12. F. Ardalan and F. Mansouri, Phys. Rev. D9, 3341(1974).
13. F. Mansouri, Phys. Rev. D8, 1159(1973).
14. Z. Koba and H. B. Nielsen, Nucl. Phys. B10, 633 (1969) and B12, 517(1969).
15. See ref. 13.
16. F. Mansouri , in preparation
17. P. G. O. Freund and F. Mansouri, Z. Phys. C14, 279 (1982).
18. F. Mansouri and L. Witten Phys. Lett. B127, 341 (1983), B140, 317 (1984), and in Proc. of VIII Johns Hopkins Workshop, ed. G. Domokos and S. Kovesi-Domokos, Word Scientific Publ., 1984.

Strings at low energy: the flux tube model

P.J.S. Watson,
Physics Department,Carleton University
Ottawa, Canada

Abstract The old string model of ten years ago has been reborn as the flux tube model. As such, it offers the best model for low energy spectroscopy and decays. This paper describes the extension to nuclear physics, showing how the properties of hadrons are modified when they are embedded in nuclear matter.

The origin of the string models lies in the spectacular success of the Veneziano model[1] as a description of meson spectroscopy and low energy (in today's language) scattering. Although this was fairly satisfactory in a descriptive sense, it ran into a number of well-publicised problems[2] in a formal sense. As a result, quark modellers turned to a variety of quasi-relativistic potential models[3] which have had considerable success without a very firm basis in a fundamental theory.These worked for isolated hadrons, but gave serious problems when dealing with interacting states. In particular, if the confining force has the form

$$V(r) \sim r^{\alpha} \qquad (1)$$

there will be a Van der Waals force of the form

$$V_{VW}(r) \sim r^{2\alpha-2} \qquad (2)$$

which is inconsistent with experiments[4]. Even if the potential has a colour interaction of the form $\lambda_i.\lambda_k$, so that there is some form of saturation for colour singlet states, there are configurations of the quarks which give rise to infinite negative energy[5].

On the more formal side, the development of lattice gauge theories have given rise to the hope that the inter-quark force could be calculated *ab initio*. The original result of K. Wilson[6] showed that the large distance force between two quarks increased linearly with the distance. Subsequent refinements have given rise to a picture where the lines of colour flux coalesce to form tubes of flux, ultimately because of the non-Abelian nature of Q.C.D. If the lattice Hamiltonian is written as

$$H = g^2/2a \; (\Sigma \; E^2(1) - 2/g^4 \; \Sigma \; (tr(UUUU)) \qquad (3)$$

where the first sum is over all links and the second over closed loops (plaquettes). The $E^2(1)$ is the energy associated with the colour electric flux along the link 1 and $U_{ik}(1)$ is a colour rotation matrix. If one demands that any physical state corresponds to operators whose colour indices contract to a colour singlet, one obtains either closed loops of flux (which can be interpreted as glueballs) or tubes which begin on a quark and end on an antiquark. Moreover in an SU(n) gauge theory, one can construct n-tube vertices, which are roughly analagous to a 3-gluon vertex in a perturbative model.

This gives rise to a very intuitive picture for hadrons, which are quark-antiquark or 3-quark systems linked by oriented tubes of flux. It offers a similar understanding of the colour-singlet nature of hadrons that a valence bond picture offers for simple chemical compounds. Furthermore, the lattice gauge theories show that at large distances the inter-quark potential has the form $V(r) = \sigma r$, while at short distances perturbation theory is good and one obtains a Colomb potential $V(r) = 4/3 \; \alpha_S/r$ and various hyperfine interactions.

This rather simple model was used by Carlson et al[7] to predict various properties of the meson and baryon spectrum. In addition to the basic flux-tube/1 gluon interaction, they use a semirelativistic waveSchrodinger equation with a kinetic energy operator of the form $H_0 = \Sigma \; (p_i^2 + m^2)^{\frac{1}{2}}$ .Furthermore, it turns out that it is neccessary to treat the one-gluon exchange in non-perturbative fashion if the π-ρ mass splitting is to be reproduced. Once this is done, the agreement with experiment is excellent.

The same basic idea has been used by Isgur, Paton and various co-workers[8,9] to study not only the meson spectrum, but also the decay widths in a simple model of flux tube breaking. An enormous amount of experimental data is explained with surprisingly good accuracy. In summary, the model provides the best dynamical quark model for low energy phenomenology. Perhaps more than anything, the applications will be limited by the lack of further experimental data, since spectroscopy is now experimentally unfashionable. The most clean-cut new predictions for the model are the pure glue states, which now correspond to oscillations of the flux tube, and not to 2 and 3 gluons confined to a bag.

It is natural to look for other applications of the same basic model. In particular, there is the attractive possibility of understanding nuclear physics from the quark model viewpoint. Of more interest, perhaps, is the possibility that new phenomena will be found in nuclear physics which require the underlying quark structure for their understanding. The E.M.C. effect[10] is a case in point: it is most simply understood by assuming that a nucleon in nuclear matter has a radius about 10% larger than a free nucleon. A second question which one should be able to answer from this viewpoint is to what extent do nucleons retain their individuality in nuclear matter: do they "dissolve", or do 6-quark complexes form?

The flux tube model is a natural framework for this problem. In this discussion, I will consider only a very simple model which can be solved analytically: a much more complicated calculation produces results differing by at most 50%. First, we use a quark-diquark approximation: from the point of view of the flux tube, the diquark behaves like an anti-quark. We suppose that a nucleon in the presence of a second may find it energetically preferable to swap quarks via a string flip-flop process. This will effectively reduce the potential between the quarks, giving rise to a reduced mass and hence a binding energy in for nucleons. The flux-tube interaction for a single nucleon

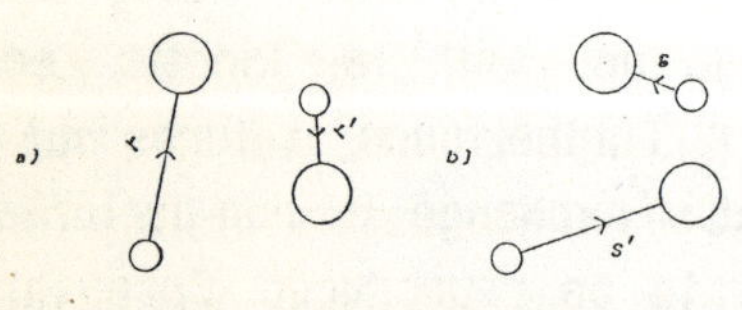

$$V(r) = \sigma\, r$$

becomes

$$V_L(r,r',s,s') = \sigma(r+r')\,\theta(-y) - \tfrac{1}{3}\sigma y\,\theta(y) \qquad (4)$$

where

$$y = r+r'-s-s'$$

for two nearby ones. If we assume a one parameter variational wavefunction of the form

$$\varphi(r) = (\eta/\pi)^{\frac{3}{4}} \exp(-\eta r^2/2) \qquad (5)$$

then the potential becomes

$$V_{eff}(r) = \sigma r - \sigma\int d^3r'd^3R\, |\Psi(R)|^2\, |\varphi(r')|^2 y\theta(y) \qquad (6)$$

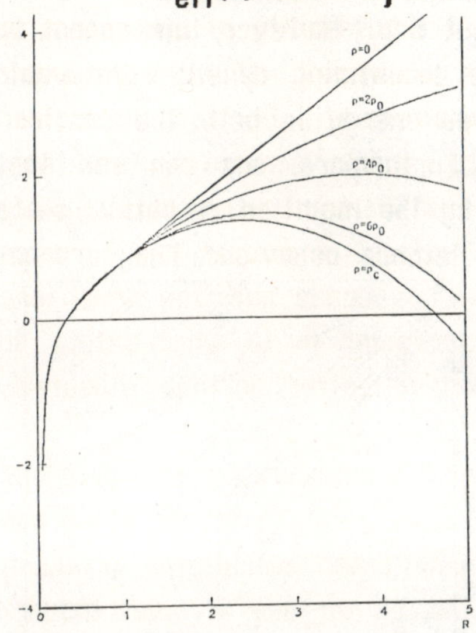

where $\Psi(R)$ is the relative wavefunction of the two nucleons (taken to be a constant $\Psi(R)=\rho$).Surprisingly, this can be evaluated exactly, giving

$$V_{eff}(r) = \sigma r - \sigma r^2\rho\pi/18\eta$$

The figure shows the results of a 10 parameter wave function: the shape of the curve is very similar. The energy is given by

$$E = 8\eta/3m + 2\sigma/(\pi\eta)^{\frac{1}{2}} - \sigma\pi\rho/12\eta^2$$

Minimising the energy gives

$$\eta^{2/3} = 4\sigma m/9 \quad (1\pm\sqrt{(1-3\pi^2\rho/4m\sigma)})$$

This actually permits one to write down an analytic expression for the binding energy of a nucleon in nuclear matter

$$\Delta E = \rho\ (3^5\pi^5/\sigma m^4)^{\frac{1}{3}}/2^6 \ \sim 8\ \text{MeV}\ (!)$$

This is a surprising result, since it depends only on σ and m, quantities measured only in high energy experiments, and the nuclear density ρ. It is only honest to point out the inadequacies of this model: we have not anti-symmetrised the wave function, and we have not included the hyperfine interaction term which gives rise to the hard-core. Hence the binding does not saturate at any ρ. Furthermore, there is no Fermi motion built into this model, so this 8 MeV should be compared to the potential energy in nuclear matter of around 32 MeV. Nevertheless, it is surprising that we obtain a result which has even a resemblance to reality.

There are a variety of other results that can be obtained from this model. From an examination of fig 3., it is obvious that at a sufficiently high density the potential will no longer support a bound state. This occurs when η from (9) becomes complex: i.e at

$$\rho = 4m\sigma/3\pi^2 \sim 3\rho_0$$

This is indicative of a deconfinement transition: it is no longer possible to define a localised wave function for the nucleon. The more sophisticated model raises this to about 6 ρ_0. However, this cannot be taken as an accurate estimate of the deconfining density; one would need to make an estimate of the free energy in both the confined (localised) and non-localised phases. Furthemore, one can see that excited states will be more affected by the modified potential, since they are more sensitive to the large distance behaviour. They are not only more tightly bound (we find approximately $\Delta E_l = \Delta E_0 \, (l+1)^{.75}$) but they also deconfine at a lower density. Even at normal nuclear densities there is a limiting l_{max} beyond which nucleons cannot exist: unfortunately $l_{max} \sim 35$, which is not capable of experimental verification. It is possible that the enhanced binding for high spin resonances could be observed by looking at strange resonances which tend to be narrower.

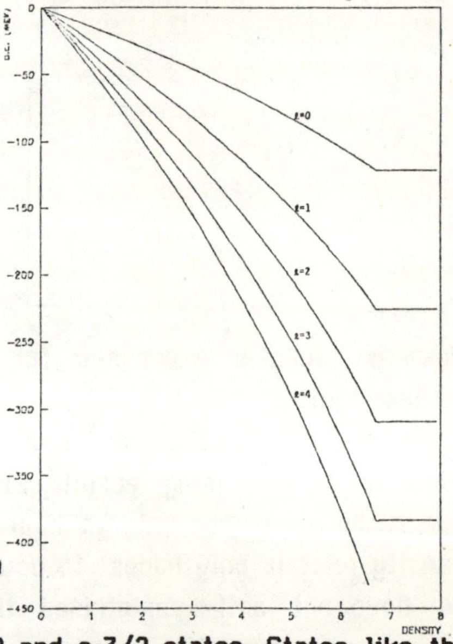

Another general observatio which is not unique to this model, is the different behaviour of s = 1/2 and s=3/2 states. States like the nucleon have a hyperfine interaction which is attractive, and in fact produces a wave function enhanced around r=0. States like the Δ_{1236} have a repulsive hyperfine interaction, which, of course, raises their mass relative to the s=3/2 states. This effect gets diluted in nuclear matter, because the effective radius increases. Hence one would expect the Δ to be much more strongly bound, via a decrease in its mass. Quantitatively the binding turns out to be about 50 MeV. The experimental situation is somewhat murky: the Δ peak in inclusive photoprduction in nuclei turns out occur at almost exactly the same energy as the free Δ. However the interpretation of this is not straight-forward, because of a variety of effects such as Pauli blocking.

It is not surprising that the increase in the radius of the bound state leads to the E.M.C. effect. This is probably universal to this kind of model. To estimate the actual size of the effect, we adapt the method of De Rujula and Martin[12]. At low Q^2,one can estimate the nucleon distribution function directly, because it is resonable to assume that it is dominated by the valence quarks. How ever the experiment measures the distribution function at a Q^2 which is many times the mass^2 of the quarks. Furthermore, it is not possible to use the evolution equations because these are only valid in the perturbative regime. However, the effects of the nuclear binding are present at large distances, whereas QCD effects are short distance effects. Hence the difference between the free and bound nucleons lies in the fact that their valence quark wave functions differ, and this difference will be maintained by the evolution equations.

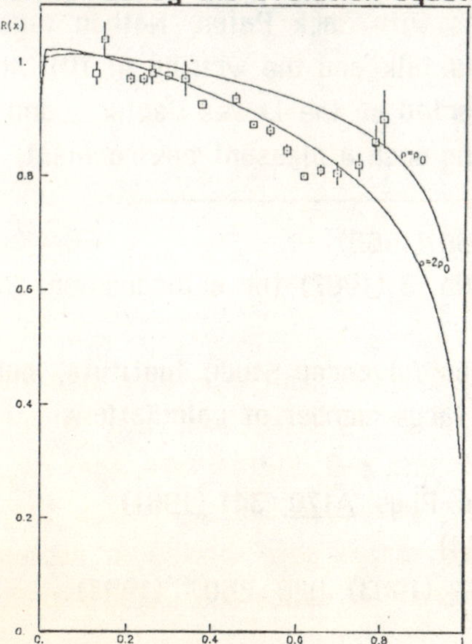

It is straight-forward to evaluate the distribution functions at low Q^2,find the moments and find R(x) using a standard parametrisation of the high Q^2 distribution function. As with all other phenomena in this model, the effect is too small; for comparaison we also show R(x) calculated at twice nuclear density, for which the agreement is good.

One final encouraging, result of the model, is that the multi-quark states (4 linked quarks in this model) turn out to give a small contribution to the binding. One can calculate the relative importance of the flip-flop and linked topologies to the binding; the latter is only about 10% of the former.

In summary, then, this model (and others like it) predicts a variety of nuclear phenomena in a fairly intuitive fashion. However, the results typically disagree with experiment by a factor of 2. It seems plausible that the results will be improved by a more realistic model, with anti-symmetrisation, hyperfine interactions and 3 quark states correctly incorporated. This last turns out to be a horrendous problem in combinatorics, because there are no less than 18 flip-flop diagrams, and a similar number of linked topologies for any random arrangement of the 6 quarks.

As a final note, it is amusing that the strings of the ten years ago have evolved in 2 very different directions. The rest of this book describes the beautiful results that follow from assuming that the underlying structure of quarks and gauge particles are strings. The model discussed here shows that these particles then interact to give rise to strings on a scale totally different from the fundamental one.

Much of this work was done in collaboration with Pierre Mathieu. I have also benefited from converstions with Jack Paton, Nathan Isgur and (in the interval between giving this talk and the writing of it!) Jim O'Connell. This work originally was started at the Leves Center , and I am grateful to Art Halprin for providing such a pleasant environment.

---

1) G. Veneziano, Nuovo Cimento 57A, 190 (1968)

2) See J. H. Schwarz, Phys. Rep. 89, No. 3 (1982) for a discussion and many references.

3) D. Gromes, Proceedings of the Yukon Advanced Study Institute, pub. World Scientific (1984) summarises a large number of calculations.

4) G. Feinberg and J. Sucher, Phys. Rev.

5) O. W. Greenberg and H. J. Lipkin, Nuc. Phys. A179, 341 (1981)

6) K. Wilson, Phys. Rev. D10, 2445,(1974)

7) J. Carlson et al., Phys. Rev. D27, 233, (1983); D28, 2807, (1983)

8) N. Isgur and J.E. Paton, Phys. Letts. 124B, 247 (1983), and Toronto preprint (1985)

9) N. Isgur and R. Kokoski, Toronto preprint, Phys. Lett. 141B, 439 (1984)

10) J.J. Aubert et al., Phyys. Lett. 123B, 275 (1983)
A. Bodek et al., Phys. Rev. Lett. 50, 1431 (1983); 51, 534, (1983)

11) P. Mathieu and P.J.S. Watson, Carleton preprint

12) A .de Rujula and F. Martin, Phys. Rev. D22, 1787 (1983)

LIST OF PARTICIPANTS

Maurice Barnhill
Physics Department
University of Delaware
Newark, DE 19716

John Breit
Physics Department
University of Pennsylvania
Philadelphia, PA 19104

Darwin Chang
Physics Department
University of Maryland
College Park, MD 20742

Wen-Kwei Cheng
Physics Department
State University of New York
Old Westbury, NY 11568

Louis Clavelli
Physics Department
University of Alabama
Tuscaloosa, AL 35486

Thomas Curtwright
Physics Department
University of Florida
Gainesville, FL 32601

Sumit Das
Theoretical Physics
Fermi National Accelerator Lab
Batavia, IL 60510

Louis Dolan
Physics Department
Rockefeller University
New York, NY 10021

Gordon N. Fleming
Physics Department
Pennsylvania State University
206 Davey Lab
University Park, PA 16802

Paul Frampton
Physics Department
University of North Carolina
Chapel Hill, NC 27514

James J. Griffin
Dept. of Physics & Astronomy
University of Maryland
College Park, MD 20742

Arthur Halprin
Physics Department
University of Delaware
Newark, DE 19716

Jeffrey Harvey
Physics Department
Princeton University
Princeton, NJ 08544

Michio Kaku
Physics Department
C.U.N.Y.
New York, NY 10031

Thomas Kephart
Physics Department
University of Florida
Gainesville, FL 32601

Samuel MacDowell
Department of Physics
Yale University
New Haven, CT 06520

William D. Mc Glinn
Physics Department
Notre Dame University
Notre Dame, IN 46556

Mark Rubin
Theoretical Physics
Fermi National Accelerator Lab
Batavia, IL 60510

Freydoon Mansouri
Department of Physics
University of Cincinnati
Cincinnati, OH 45221

Kunnat Sebastian
Physics Department
University of Lowell
Lowell, MA 01854

James O'Connell
Center for Radiation Research
National Bureau of Standards
RADP C229A 02115
Washington, DC 20234

Floyd Stecker
Lab for High Energy Astrophysics
NASA Goddard Space Flight Center
Code 665
Greenbelt, MD 20711

Palash Pal
Physics Department
University of Maryland
College Park, MD 20742

Peter Watson
Department of Physics
Carleton Universtiy
Ottawa, Ontario K1S 5B6
CANADA

Stuart Raby
Los Alamos National Laboratory
T-8
Los Alamos, NM 87545

P. K. Williams
Div. of High Energy Physics
ER 22 MS G256
Department of Energy
Washington, DC 20545

Norman Redlich
Department of Physics
Brandeis University
Waltham, MA 02254

Alexander Zaks
Physics Department
Northeastern University
Boston, MA 02115